Jesus: In Ethiopia

The Meeting Between The FATHER and The SON

Let the Adventure Begins

Jim Rankin

Jim Rankin
ISA. 40:31

Copyright © 2011 by Jim Rankin

Jesus: In Ethiopia
The Meeting Between The FATHER and The SON
by Jim Rankin

Printed in the United States of America

ISBN 9781619046177

All rights reserved solely by the author. The author guarantees all contents are original and do not infringe upon the legal rights of any other person or work. No part of this book may be reproduced in any form without the permission of the author. The views expressed in this book are not necessarily those of the publisher.

Unless otherwise indicated, Bible quotations are taken from The Kings James Version.

Cover photograph
Copyright © 2011 by Jim Rankin, All rights reserved

Edited by Sue Bronner

www.xulonpress.com

To Sherri and our incredible children as we continue our God led journey through life.

Contents

Acknowledgements ix

Foreword xi

Prologue xvii

Chapter 1: On Holy Ground 23

Chapter 2: Let the Adventure Begin 37

Chapter 3: God Arranged Meeting 55

Chapter 4: The Bethlehem Calling 65

Chapter 5: A Nile Journey 74

Chapter 6: To the Land Beyond 94

Chapter 7: Beliefs and Writings 118

Chapter 8: To the Banks of a Mighty Lake 137

Chapter 9: The Lost Island: Host to a
Heavenly Summit 158

Chapter 10: Along the Path to Tana Kirkos.......................190

Chapter 11: Ancient Writings: Jesus and
            The Father Meet 204
Chapter 12: So Why Ethiopia? 230
Chapter 13: Putting the Puzzle Together 256
Chapter 14: Were They Really Lost Years? 286
Chapter 15: Our Adventure Begins Again 303

ACKNOWLEDGEMENT

This project has been part of an amazing adventure in my life. It has been a journey that has been full of intrigue, amazement, revelation and many tears in order to complete. Stepping back and looking at it all now, I have come to realize that it would not have been possible without the assistance of many individuals along the path of my life. There are so many that have played an intricate part in preparing me for this trek that it would take another book to thank them all. I will just say that I regret that I am unable to share each of their names within this acknowledgement, but I want them all to know that their reward is yet to come.

I must extend thanks to a few who really have trusted me to not only accomplish this life-changing task, but also to meet the challenges that have been given to me throughout my life. To these people I owe my most humble gratitude:

my wife Sherri for her unmatched love and encouragement in every victory and defeat; my wonderful children (Shayne who is with the Lord and rooting for me) and Austin and Skyler for their incredible endurance to stand behind me at any cost; my parents (Ronald who is also with the Lord) and Betty for their belief in me from the very beginning; Dr. and Mrs. Verlis Collins for their prayers and trust in my ability to reach the world for Christ; Bob Cornuke for the following of his heart and his teachings to guide me on this journey, and my Lord and Savior Jesus Christ for His sacrifice for me.

Because of these people and many others, my life has been filled with prayer, support, and guidance along every journey I have already taken. But the adventure isn't over, and I now thank those I haven't yet met who will be a part of future explorations. This journey is only the beginning, and God only knows where His next exploration will lead, but I do have the surety that wherever He leads, I will follow.

FORWARD

By Robert Cornuke, The Legend Chaser

As we journey through life we encounter many types of people. As a former police investigator, I came across both the gentle and the notorious. Later I came in contact with Jim Irwin, the eighth man to walk on the moon, and his sweet spirit led me to join up with him to begin a quest for God's truth in the Bible with our search starting with the Ark of Noah.

Since then, my encounters with people have taken me to the farthest extremes; from my search for Mount Sinai in Saudi Arabia I encountered radical extremist and was put under arrest many times as I strived further into my quest for the truth. Then in the late nineties I started my research for the Ark of the Covenant traveling from Jerusalem to Ethiopia.

In that research I have been able to piece together an incredible amount of evidence, as seen in my book *Relic Quest*, to back up my claims that are powerful enough to stand up in a court case that the Ark may still reside in Ethiopia today. It was during this quest that God led me on another adventure after I encountered a shipwreck in Lake Tana, near Barhir Dar, Ethiopia. This mishap led me to search for the actual shipwreck of the Apostle Paul. But while in Axum, Ethiopia I not only continued my research, but because of their undying faith I have since developed a passion for these amazing people and have returned to Ethiopia in some cases several times a year. These returns are not only to continue my research, or to host a documentary, but to do whatever I can to help these hurting people of this beautiful land.

The funny thing is you never know what or who God is going to lay in your path in life. Although I have had my share of strange encounters and met some of the most interesting people over my journey through life, there was no foretelling of the meeting that I would have with Jim Rankin and what it would mean for our future explorations in Ethiopia. A phone call to our organization, The BASE Institute, to order the DVD of *Mountain of Fire: The Search for Mount Sinai*, that he remembered airing on the History Channel, would turn into a long lasting relationship with biblical proportions. I remember that moment: as I was preparing to leave for Malta to film a documentary on my find of the anchors of Paul's shipwreck; this young man called and wanted the Mount Sinai DVD. He struck me with his poise and demeanor, with a knowledge and desire to learn more. We all want adventure in our lives, but this guy seemed to have adventure following him.

After returning from Malta, I couldn't get his name out of my mind as a perfect candidate to return with me and our team to Ethiopia to continue our research for the Ark of the Covenant. So, I called him one afternoon and told him I felt he needed to be with me on our next journey to this ancient land in Africa. After some thought and prayer, he and his wife accepted my offer and I'm glad they did. You never know how God is going to move in your life and I have seen Him reveal unbelievable truths to me during my explorations all over this world. He has directed me from harm's way and led me to ancient biblical truths that I would have never been able to solve without His blessings guiding me along each trail.

During Jim's travels with me, I had the opportunity to see God's revelation on him in Ethiopia. His kind and witty spirit keep you on the edge of your seat. His knowledge of God's Word brings him to tears and it's evident that he is being led by the Lord in many ways. But it was his desire to know and search the truth that struck me. It reminded me a great deal of myself, when I stepped back and tried to figure him out.

I watched as bread crumbs were being laid in his path in Axum as he shared his "Arise and go" theory with our exploration team and how the artifacts in the treasury seemed to be calling his name. He didn't stop with the status quo, he wanted more insight, more facts and I watched him turn to the Bible to confirm the answers he was receiving. When we arrived at one of the many small airports during our journey, I remember him grabbing his bag and walking by me on the tarmac and saying, "Bob Cornuke, you and I are much more alike than you know, my friend." At the time I had no idea what that meant, but since then I have seen much in him that has directed me throughout many of my adventures thus far.

Never will I forget his time on Tana Kirkos Island, though. When he knelt in the spot the Abba claimed to be the holiest spot on the island, it was like you could see a change coming over Jim. Tears began to flow and the wheels in his head began to spin in high gear. It was God's revelation to him of something that he was to share with the world. This book is a confirmation of that. Just as I had done in my search of Mount Sinai, Paul's shipwreck in Malta, and with the Ark of the Covenant, I was witnessing someone else putting the pieces together that the Lord laid out perfectly in front of him. What impressed me more is that he did not stop there. His research in this book is phenomenal and the hand of God leading him along this journey is without question, heart pounding.

Since our first journey to the ancient land of Ethiopia, Jim has teamed up with me to continue the research for the Ark of the Covenant and to share the Gospel with the people there. He even had Scriptures translated into their Amharic language in order to open up a whole new window of faith in Christ during our explorations in their land. Never dismiss the people that you come in contact with because you never know what blessings are in store for you.

This book is the ending, or maybe the beginning to an incredible claim that the Ethiopians have kept under lock and key for centuries. You will be enlightened, blessed, and revived in not only the truth, but also the faith of hearing God's calling and someone following where He leads in this writing. I'm thankful for our accidental phone meeting, and more thankful that we will be joining together on many more adventures to come. I guess the words he spoke on the tarmac are true, "…you and I are much more alike than you know, my friend." Yes, we are friends with many more explorations ahead calling us both.

Prologue

The extreme heat of the desert is a very unfriendly place to anyone, especially for a young family traveling alone and on the run. The scorching sands can feel like coarse sandpaper scraping on open flesh in the winds. The blaring heat can reach temperatures that, within minutes can blister exposed skin. The desert can dehydrate a body in a short time causing a miserable and sure death if ill prepared.

The desert is where this story begins to unfold. This young family was no ordinary family, though. They were sure to have limited fresh water supplies making their throats heave with convulsions as the scorching sun sandblasted their lips in layers. As beautiful as the desert may appear, it can be a quick end to anyone attempting to travel through it un-experienced. Newly discovered historic writings reveal that this family also endured a danger of another sort as they traveled through crocodile and hippopotamus-infested waters to reach their necessary and prophetic destination. Some 2,000 years ago a young man, his wife, her cousin and their young Son began this journey, wrapped in mystery, speculation, and wonder for most that have searched the Scriptures for the early years of Jesus' walk on this earth. When His earthly father, Joseph, was confronted by an angel and instructed to take the young boy and His mother, Mary, to Egypt to escape the bloodthirsty killings of Kind Herod's army, most have never followed this Holy family's journey any farther. Were His travels into Egypt uneventful and was the trek into the desert and southward an unimportant timeline in the life of the Savior of the world? According to the book of John in chapter 21 we read, *"And there are also many other things which Jesus did, the which, if they should be written every one, I suppose that even the world itself could not contain the books that should be written. Amen."* In just a very few words the Apostle John gives a wide open opportunity to understand that there was much more to the story of Jesus Christ's walk on this earth. John simply confirms that the Bible gives us what we need to know for the instruction and teachings of Jesus. But, just as we record the history of George Washington and Abraham Lincoln, so has the life and times of Jesus' unmentioned biblical biography been logged in scrolls and writings, hidden away until the appointed time that God feels that it needs to be revealed. John relates that much more to the story of Jesus was never written down in the Gospels, but the stories were engraved on the pages of lost documents locked away from the curious eyes of the world.

This is an open addition to the biography of Jesus describing a lost meeting between the Son and the Father. That is, lost until now. Perhaps traditional church leaders and scholars will not accept the writings of this book or this newly found information. Some of those who criticize might never take the time to venture beyond the conventional wisdom and tradition. That's why I decided to explore beyond the familiarity of my comfort zone and by taking God's lead and letting Him remove me from a world I know and into a world that has long been forgotten. Since God's calling is not without purpose, He often leads us into life changing revelation.

There have been many who have taken the risks needed to follow the lead of the Lord. Bob Cornuke is one of those risk takers. As a former police investigator turned biblical explorer, his relationship with astronaut Jim Irwin started his quest by assisting in the search for Noah's Ark. Since that time, Cornuke has been credited with discovering the authentic Mount Sinai, the Shipwreck of Paul and his incredible discoveries with the Ark of the Covenant's possible location. The late Ron Wyatt is another risk-taking explorer, a nurse anesthetist from Nashville, Tennessee who was credited with the find of Noah's Ark, although this project is still under investigation. Risk-takers Jim and Penny Caldwell, while working for an oil company in Saudi Arabia searched the site of the real Mount Sinai and discovered the split rock that Moses struck to supply water to the masses. Where they found this incredible monolithic stone enshrinement, marking the place where an amazing miracle of the past occurred, is nearby the exact location that Cornuke and Larry Williams confirmed as the true Mount Sinai. These are just a few who have risked much to follow God's lead to bring us unbelievable revelations that the Lord has decided to manifest to the world. This is exactly what happened to us as we disregarded traditional acceptance of man's theories and allowed the Lord to lead us into an unknown and somewhat uncivilized corner of the earth to discover something of the early travels of the Holy Family.

So what did happen to the Boy Savior during His family's retreat to Egypt? Was it uneventful? Or, was it a set-up for Jesus to prepare for the rest of His life on this earth? There have been many publications and speculations about the whereabouts of the Holy family, Jesus the boy, Mary and Joseph, in Egypt. The other question that begs to be asked, "But where else did they go?" Was Egypt the only place that the Holy family traveled during this time? The answer to this mystery has just become an assumed factor for hundreds of years, and this is where our journey begins. We start in the ancient land of Israel in the little town of Bethlehem. We'll then take you to Egypt and walk alongside the Holy Family as they struggled through the heat and sand, escaping Herod's soldiers and facing angry mobs and thieves. We'll look at the historical writings that provide information of Jesus' whereabouts during His family's retreat into Egypt. Or, was it a retreat after all? The Scriptures only tell us that the Holy family was told to go to Egypt; it doesn't tell us this is the only place where they ventured. As a matter of fact, that's the question that baffled me. Why were they told to go to Egypt when there are plenty of surrounding lands they could have found refuge in? Following the ancient Coptic and historical writings however, you start to see a new pattern, a pattern that led the fleeing family to a faithful meeting with God the Father and His Son, Jesus.

Returning to these ancient writings and following God's lead, we took the reins of camels, bounced along desert paths, and discovered people and resources who presented us with a long lost story. We scaled cliffs, traveled to mysterious islands and lowered ourselves into tombs. We partook in the successful translation of a nearly extinct and remotely used ancient language to bring you this miraculous story of discovery. So, we invite you to mount up the camels and join us as we begin our travels through the Egyptian deserts and along the Nile toward the south into the land of Ethiopia. South is the key to our journey as we travel to the far reaches of the legendary land of Ethiopia to unveil the mystery of a long lost meeting between the Father and the Son. Let's open up the pages and journey to the forbidden island with Jesus: In Ethiopia.

Chapter 1

"ON HOLY GROUND"

Nerves had to be tightened as the skies darkened over the mountain top. The mountain blazed as if it were on fire, glowing in the night and engulfed in clouds mysteriously consuming it. It was the Holy mountain of God, Mount Sinai. This mountain was so holy that no one stepped foot on it unless called upon by God Himself to walk upon these ordained slopes. One day long ago Moses, a lone individual, chosen by the Lord to make the trek, had no idea in his mind what he was about to encounter or embark upon. A voice commanded, *"put off thy shoes from off thy feet, for the place whereon thou standest is holy ground"* (Exodus 3:5). One can only imagine the whirlwind of thoughts that must have spun in Moses' head as he apprehensively approached a most frightening sight. Before him, as he came closer to the Voice that uttered those words atop the burnt mountain, he saw a bush burning brightly, though not being consumed, and the Voice of Jehovah God barreling out His instruction and commands upon the one who was chosen to lead the enslaved people. He was to lead these people who had been held in bondage in Egypt back to the Holy Land. He would lead them back to the precious Lord-ordained land of milk and honey. The thought of hearing those words, *"for the place whereon thou standest is holy ground,"* must have brought Moses to a gut wrenching shivering halt as he quickly removed his sandals, realizing that he was in the presence of God Almighty.

Moses had an enormous task on his shoulders because no one would have ever imagined that Moses would have been "the one" chosen to fulfill this command. Having been cast out from the land of Egypt as a once thought prince to his speech impediment that would surely hamper relating the message from God, we see the continued pattern move mightily in the tradition of the Lord to choose the most unlikely to complete the most miraculous. Moses' humble beginnings bring him from a family of Hebrew slaves, to the royalty of the Pharaoh's palace, to the "holy ground" on the top of Mount Sinai. In his own right, Moses took quite a journey culturally, emotionally and spiritually to reach the point that God had entrusted in him to carry out His plans. That phrase will ring true, *"for the place whereon thou standest is holy ground,"* as we continue our own modern-day adventure as well. As a matter of fact, it will be a phrase that will send shivers down my spine as we travel throughout this journey. Far before I became part of this exploration, I had begun writing a play for my church called "On Holy Ground." It was to be a dramatization of Moses' time on Mount Sinai and his conversations with God during that time. One can only imagine some of the discussions that took place. But, later in our journey you will see why this phrase came to a startling truth for me personally. This phrase will reveal itself as one of the keys that God was speaking to us through impressions that He will lay in our thoughts and on our paths. Throughout this adventure we will see this happen several times as we are guided to the destination that we have been sent to discover.

Let's start this story from the time Moses came down from the mountain from *"Holy Ground"* with God Himself. Following the biblical flight-line, the traditional location of the "Helena" site of Mount Sinai on Jabal Musa in the Sinai Peninsula, God's Holy Mountain, in no way matches any account shown clearly in the Scriptures. At the base of this mountain is Saint Catherine's Monastery along with a collection of theme park-like souvenir hounds offering their wares to gullible tourists looking for anything relating to their visit to the mountain. Years earlier my brother-in-law even made the trip to this location during a stint in the military and fell to the belief and frenzy that this was the Holy mountain of God. Tourists and pilgrims for centuries have been lured into this belief and, in some odd way, have come away fulfilled with a false fullness of the spirit within.

Helena, the mother of the Roman Emperor Constantine, chose this location as Mount Sinai. Helena ventured out with a strong coalition of soldiers in pursuit of artifacts and biblical locations, as she felt her calling to be. In many cases, force was exerted to achieve this purpose. With many of the Helena claims, there are no factual or logical trails of substance to bring reliable truth to many of these sites. Regardless, her unsubstantiated claim that Jabal Musa, also known as Mount Horeb, has been the traditional site drawing millions of worshippers throughout the centuries paying homage to more of their faith than with the location. Even biblical scholars aren't completely sold on the idea that this is the true Mount Sinai. You will find this as fact when examine biblical maps found in most everyday Bibles. Go ahead, take a look. In most Bibles the maps will indicate a question mark next to the location of Mount Sinai. That leads us, or should lead us, to look further and to discover a more accurate biblical description.

The most likely site is reported by biblical explorer Bob Cornuke, and is located in Saudi Arabia, just as the Apostle Paul accurately distinguishes in the book of Galatians 4:25; *"For this Agar is mount Sinai in Arabia."* Amazingly, from the natural land bridge in the Red Sea that the Israelites would have crossed (Exodus 14 and 15), to the bitter springs of Marah (Exodus 15:23), to the twelve wells, or springs, and seventy palms of Elim (Exodus 15:27), to the mount itself, Jabal Alaz, is near evidence enough. But, if you are actually able to reach this restricted Saudi area, an off-limits military site and archeological area, you will see warning signs that trespassers will be prosecuted based on an antiquities law of the country; however you would also find enough evidence to stand up in a court case. Cornuke and his friend and colleague Larry Williams (a very successful commodities investor, former politician and explorer) discovered a stone altar engraved with Egyptian cows similar to those of the Egyptian gods Hathor and Apis, as found in Exodus 32. Of course, this Scripture also states that the people ordered

Aaron to shape a golden calf and place it on the altar they had built. As in Scripture, we notice that there was more than one calf noting the many calves engraved in the altar in Saudi Arabia and matching Exodus 32:4, *"after he had made it a molten calf: and they said, These be thy gods, O Israel,."* That explains the altar for the golden calf in addition the calves etched into the stone itself.

Bob and Larry locate the barriers around the mountain described in Exodus 19:12:

"And thou shalt set bounds unto the people round about, saying, Take heed to yourselves, that ye go not up into the mount, or touch the border of it: whosoever toucheth the mount shall be surely put to death."

The two explorers also came upon twelve columns and the sacrificial altar as we read in Exodus 24:4:

"And Moses wrote all the words of the LORD, and rose up early in the morning, and builded an altar under the hill, and twelve pillars, according to the twelve tribes of Israel."

As shared in the Scriptures, a cave is located in the face of the mountain where Elijah took shelter on Mount Sinai in 1 Kings 19:13:

"when Elijah heard it, that he wrapped his face in his mantle, and went out, and stood in the entering in of the cave."

But, one of the most stunning features is the coal black rock that only this mountain holds. It's a very unusual black rock that when broken open shares the same rust orange colors of the rest of the rock common to the area. This coal black covering may help explain the Scripture in Exodus 19:16-18:

"And it came to pass on the third day in the morning, that there were thunders and lightnings, and a thick cloud upon the mount, and the voice of the trumpet exceeding loud; so that all the people that was in the camp trembled. And Moses brought forth the people out of the camp to meet with God; and they stood at the nether part of the mount. And mount Sinai was altogether on a smoke, because the LORD descended upon it in fire: and the smoke thereof ascended

as the smoke of a furnace, and the whole mount quaked greatly."

Then, with Jim and Penny Caldwell's find of Moses' split rock nearby, it puts the final stamp on the case. The evidence that the Caldwell family discovered while at the split rock, including the obvious dried water trail leading from the rock into the valley and the flaked erosion from the bottom up giving the overall impression that the water was gushing from the base of this huge stone monolith, completely brings to life the incredible find of Cornuke and Williams at Mount Sinai. This rock was split when Moses took his staff and slammed it against the rock causing water to gush out into the valley below as seen in Numbers 20:7-11.

"And the LORD spake unto Moses, saying, Take the rod, and gather thou the assembly together, thou, and Aaron thy brother, and speak ye unto the rock before their eyes; and it shall give forth his water, and thou shalt bring forth to them water out of the rock: so thou shalt give the congregation and their beasts drink. And Moses took the rod from before the LORD, as he commanded him. And Moses and Aaron gathered the congregation together before the rock, and he said

unto them, Hear now, ye rebels; must we fetch you water out of this rock? And Moses lifted up his hand, and with his rod he smote the rock twice: and the water came out abundantly, and the congregation drank, and their beasts also."

With this discovery we have to rely more strongly on the evidence than with the tradition, that Moses actually went up to meet with God on holy ground and then descended back down to his people carrying the Ten Commandments of the Lord Himself. It was at this place that God commanded Moses and his brother Aaron to construct a box out of acacia wood, or shittim wood, and then overlay it with gold to contain the commandments, and then to build a solid gold lid with two cherubim angels atop with their wings folded and facing the center point of the lid known as the Mercy Seat of God. This would be the place where God would dwell upon His appearance on the Day of Atonement in the temple or tabernacle. Along with that, it would also serve as God's Holy throne upon His arrival on this earth.

As Aaron, Moses' brother, became his spokesperson, and eventually the high priest to carry out the meeting with the Lord in the tabernacle during the trek to the Holy Land, the Ark of the Covenant became possibly the greatest and most mysterious artifact in history. Its power was unstoppable and the miracles that surrounded it were beyond human understanding. From bringing down the massive stone walls of Jericho to parting waters of the flooded Jordan River to defeating armies, the Ark of the Covenant was a holy artifact that was worshipped by all who believed in God and feared by everyone during the time of Moses and even today for many. Even to touch the Ark would bring death because of its holiness and the separation between God and sin. Even now a replica of the Ark of the Covenant is an object that is highly revered and thought to be a holy artifact because of what it represents to many faiths. In Ethiopia you will find a replica of the Ark, or a tabot, in every church inside a Holy of Holies. If, for any reason, even the replica would be removed from that church it would literally mean that the church would lose its holiness and become deconsecrated as a result.

In the ancient times, on the Day of Atonement the high priest would go to the Tabernacle to prepare himself before entering the tented structure. The priest first cleansed his body of sin and then went into prayer for the forgiveness of sin of the people. It was a literal spiritual removal of the sin before he presented himself in front of God in the Holy of

Holies. This same ritual continued into the time of Solomon's temple and even up to the time of Jesus' ministry. The high priest would have to physically wash himself completely and then he would be dressed into seven holy garments which was representative of honoring Christ. He would wear a long coat, linen breeches, and the robe of the ephod colored blue with golden bells sewn into the hem. The most important part of his dress was the ephod with the outer material made of fine twined linen representing Christ's purity, with gold; representing Christ's deity, with purple; representing Christ's royalty, with blue; representing heaven, with scarlet, which showed the coming sacrifice of Christ. On the front it contained, the stones of the twelve tribes of Israel which represented the burden of the people carried on the shoulders of the high priest, just as Christ carries our burden on Himself. Then there was the girdle, the breastplate and the turban, or mitre, which he would wear on his head.

Properly attired the high priest would then enter the Tabernacle with the bowl of sacrificial blood and pause in the first chamber before the table of shewbread, altar of incense, and the golden candlestick lamp-stand. He would prepare himself of purity once again, as he did with the outer cleansing before entering the Holy of Holies. Inside the

Holy of Holies, the priest would again go into a solemn issuance of prayer and repentance for himself and the people, and then sprinkle the blood of the sacrifice directly on the Mercy Seat for cleansing where God Himself would sit, and then on the floor beneath for cleansing where the souls of God's feet would be planted. God's desire since the Garden of Eden was to dwell with us, but our sin has kept Him from us. That's why so much care had to be taken to cleanse even this small area for the arrival of the Lord Himself. If for any reason God didn't accept the plea of the high priest for forgiveness, the priest would be struck down by the power of God through the Ark. He would then be pulled out by the rope that was attached to him if the people didn't hear the bells jingling anymore from inside the tabernacle. If all went as planned, God would forgive His people and bless them. Finally, following the beginning rituals, God would then give additional instruction for His people or to prepare them for whatever was to come.

The Ark of the Covenant would take its rest inside the Tabernacle tent along the journey of Moses and the people of Israel on their trek to the Holy Land. Eventually Solomon constructed the temple in Jerusalem that was thought to be a permanent home to the Ark of the Covenant. Sin, defilement, and destruction of man, however, eventually put an end to that original plan. The Holy of Holies was considered so deified that only the specifications that God laid out could be housed within that place. Anything else would desecrate the room and God's presence would be removed. The Levites were extremely careful and protective over the Ark and the Holy of Holies in order to carry out their appointment as the overseers of this Holy object.

Just as God told Moses to remove his sandals when he was walking on Holy ground on Mount Sinai in the presence of the God the same was true for the high priest as they entered into the Holy of Holies. There was nowhere that projected the divine nature of God's presence on this sinful earth than in that chamber where the Ark of the Covenant rested and between the Cherubim on the Mercy Seat, which was literally the throne of God. The Ark of the Covenant and the Mercy Seat carried with them heavenly force not to be reckoned with.

From this point, the story of the Ark in the Scriptures is far from over. The sacrificial need for the Day of Atonement and the dwelling of God on the Mercy Seat continued until the Ark of the New Covenant in Christ was brought to this earth. The sacrificial ceremonies were a necessity throughout the ages. Then, as we know, Jesus became the Law and carried them within Him, thus becoming the sacrificial Ark. The fact is, "Holy Ground" plays a divine role throughout the ages. From the Garden of Eden, where God dwelled among Adam and Eve, to Mount Sinai, where Moses knelt in humble fashion on Holy Ground, to the Holy of Holies in the Tabernacles and in the Temples themselves, God's desire to be with us has never ceased. That's why those words "on Holy Ground" play such a prominent role in our journey and in our everyday lives in search of the truth of the meeting between The Father and The Son.

Chapter 2

LET THE ADVENTURE BEGIN

Looking back, the arrival in Ohio was a bit of a shock to this Central Floridian. With my dad retiring from the Air Force, he wanted to return to his home in Southwest Ohio to where he began his journey twenty years earlier at Wright Patterson Air Force Base in Dayton, Ohio. Knowing that the family would have to make their way on ahead, we began our journey northward months before he would be able to finally retire. Within a year, as I had just turned six years of age, the family on my father's side had begun an annual gathering, or reunion, in the backwoods area of Southwest Ohio, far from the city, into the hilly woodlands. This was a land I wasn't familiar with and to be quite honest, not too sure of either. There were no swamps with alligators, the people talked differently, their accent was unfamiliar, and they dressed a little weird according to Florida standards, but that's a whole different story and maybe one to explore at another time. When we arrived at this back hills country setting at the end of a dirt and gravel road, we would begin the rugged drive back behind my grandmother's old mobile home.

One of my earliest recollections of my grandmother, who was well into her eighties, was her sharing stories of being a little girl in the late 1800s and jumping into the twentieth century. She was a kind woman but very strong willed in her age. On our very first day in her small trailer, my uncle had caught a huge snapping turtle in a nearby pond; my grandmother, cane in her left hand and a sharp machete in the other, stepped onto the turtle's back and with one whack, took off his head. I'll never forget her saying to my dad and uncle, "Skin him and bring him in to boil." I could tell quickly that she was a survivor, and I knew this woman was someone I wanted to know better. This was a woman who had been through the struggles of life and had the battle scars to prove it. But, at the same time, she had a passion to share her life with the ones she loved. Stories of seeing President William McKinley, the 25th President of the United States and attending the Buffalo Bill Wild West

Show proved this woman had even more stories to keep a young explorer like me wrapped up for days. One of her most intriguing stories that she shared with me was about the contents of an old wooden box containing the keys to a trap door at the home Reverend John Arthur Rankin, one of the initiators of the Underground Railroad, in Ripley, Ohio which included a collection of documents that he issued to those who made their way across the Ohio River to proclaim them as free slaves. Unfortunately, after the passing of my grandmother, the box and its contents disappeared, another example of neglected history.

After visiting my grandmother, we drove on this worn and rutted driveway into the back country to a nearly washed away dirt path into a wooded area where the families would gather and pitch their tents, build up the wood for an evening fire, and set up the card tables for endless tournaments of euchre. For this adventurous young boy, I wanted nothing to do with that. There was wilderness to explore and the boring thought of planting myself at a euchre table was not part of my adventure. They played their game all day long, it seemed like to me, or, as I would hear often, "until the cows come home." So, in the midst of their entertainment, I would disappear into the forest and not return until it was time for the late evening dinners. A growing boy had to eat, and I wasn't about to miss the evening meals which always consisted of homemade fried chicken, mashed potatoes, and usually corn bread. It makes me hungry to recall that delicious childhood memory.

During my adventures I would swing on vines, slide down shale covered embankments, and build forts to hide away in. I would pretend I was a combination of Tarzan, Daniel Boone, Jim Fowler from Mutual of Omaha's Wild Kingdom, Frank Buck the early movie adventurer, and Evel Knievel all wrapped into one young boy. Maybe even thoughts of Superman, Batman and Captain America would be a part of my day. I was always open to adventure, and I had the imagination to create whatever the surroundings presented me with. One afternoon, while sitting near the creek bed up against an oak tree, I started digging in the dirt and saw a series of little stones that appeared to be fossilized Cheerios laying arranged in a perfect circle. To my surprise, there were dozens of them in a pattern right next to the tree. As I kept digging I came across an arrowhead. Not just one, but four of them all in the same area as if they had been buried there together in something that had rotted away over the centuries. Quickly, I began to realize that these weren't ancient Cheerios, but some type of bead or fossils carved as beads for ceremonial and decorative dress for ancient Indians.

With this new found discovery, I decided to try my luck in another area about fifty yards from this particular spot where two creeks forced into one bigger one. Again, Indian beads, arrowheads and even an ax head this time. Had I by accident stumbled across some type of former ancient American Indian village from centuries earlier? Over the next six years I would fill coffee cans with beads of all different sizes, arrowheads so perfect that they would still take down a large deer, ax heads, and even staff heads and pottery. Eventually it became obvious I had been excavating an ancient Indian village that had been lost for centuries. My brothers didn't have much of an interest in the excavation, but occasionally, my sister Debbie would join me in the hunt. But, after my grandmother passed away, the reunions in this place would gradually dwindle, and I felt it best that this area remain untouched until some other day. As a matter of fact, nearly fifteen years later I would bring my own children to this site to explore and dig up some of those incredible artifacts. Still today, this remote area in Ohio remains the way it was when I was a boy. As for me, other than crabbing off the beaches of Florida and enticing alligators in the swamps, and now exploring and discovering an ancient Indian village in the woodlands, this need for adventure and the desire to explore has never gone away.

Needless to say, my life has been a wild adventure fulfilling nearly every dream and passion that I have ever had. I've met and worked with some of the most noted celebrities over the ages while traveling from coast to coast in my professional pursuit of the thrills of life. But, the desire to explore would never leave me. I've shared the stage with some of the entertainment business' top performers; stood in the presence of presidents; jumped a few motorcycles and built a great relationship with notables as Evel Knievel, Carl Perkins, and many others. Despite these thrills, my dream for historical adventure was still my unrequited passion. This brings me to the not too distant past.

A Family Getaway With A Twist

It was a steamy morning in late July 2009. My family and I had made a quick getaway from Ohio to the Great Smoky Mountains in Eastern Tennessee. One of my specific rules when taking road trips is to make very few restroom stops so we can reach our destination and enjoy our time there, rather than spending extended time on the road. So, in order to keep the thought of having to stop out of my wife's mind, she would bring pillows and blankets and nestle into her seat and sleep for most of our trips. When we traveled to Tennessee, everybody knew that our first stop would be at the Tennessee Welcome Center at the border with Kentucky. From that point, we didn't stop again until reaching the Smokey Mountains which was about an hour and a half ahead. After our five hour journey making good time, we finally took the exit off of Interstate 40 to Sevierville, Tennessee. After driving through the home town of Dolly Parton and checking out the sites in Pigeon Forge, we then took the short wooded roadway to Gatlinburg, our home for the next three days.

When we arrived, the town of Gatlinburg was bustling as always. The traffic was backed up as anxious tourists made their way into town to seek for that lost treasure in one of the hundreds of shops, to visit the aquarium, to head up the mountains for a quiet drive, or to find a place of refuge in during their time there. We pulled down a side street called River Road, which ran parallel to the Little Pigeon River. Turning right and crossing the concrete bridge over the river to our hotel, we had finally arrived. After checking into the

Edgewater Hotel, the tradition has always been to get out and check the sights and sounds and to explore what's new and what's old. It always seems like the same thing, but it wasn't home, so it was exciting to explore the "strip," the downtown area of Gatlinburg. Once we made our way back to the hotel, we enjoyed a swim and a relaxing splash in the hot tub before finally retiring for the night. I was beat, tired from the earlier drive and worn out from walking up the Gatlinburg strip. The heat and humidity had physically drained me, and it was time to call it a night. It didn't take long for my eyes to close and my mind to lapse into a state of dreams.

Within a short time into my calm sleepy slumber, I had a startling wake up around 3:00 a.m. that sent me flying straight up into a "catch my breath" upright position. My wife just sighed and turned the other way oblivious to the dramatic event that was about to reshape our lives. I heard a voice, a whisper, or even a thought that came over me for the need to research the Ark of the Covenant of the Old Testament. Nothing else, just the Ark of the Covenant! Sure, I had knowledge of the Holy Ark of God, but never had I felt compelled to research it more fully. After all, my favorite film of all time was *Raiders of the Lost Ark*, and I thought

I was Indiana Jones himself, or at least that's what my wife would call me when I would take off on some far-fetched adventure. But, this awakening was the simple thought of the Ark of the Covenant and nothing more. I was clueless on this one but knew that there had to be some meaning behind it; it was too plain not to have some purpose beyond my understanding. I lay my head back down deep in thought before finally dozing off to sleep.

After we awoke a few hours later, I walked out on the balcony to take in a breath of fresh morning mountain air. I could tell right away that it was going to be another steamy day that was common in this area. We loaded up the family and left Gatlinburg for Pigeon Forge and joined thousands of others on a tram to Dollywood Amusement Park. That's when this whole journey began. My cell phone started to ring and my first thought was "should I answer or not?" I'm trying to get away with my family, and I didn't feel like fielding any phone calls. To the dismay of my family, I did it anyway. It was someone claiming to be an associate with the *Twilight* films and he was in Vancouver on production of a new film to be released in June 2010. I half heartedly thought it was one of my friends playing a feeble joke on me, and hung up. Fifteen minutes later, as I stood in line for the

Mystery Mine rollercoaster with my sons Austin and Skyler, my phone rang again. I hesitantly answered. It was the same man again. "I'm not kidding," he said, "I'm with the production of *Eclipse*, the third of the *Twilight* Saga films. We're feeling a little 'Crabby' in Vancouver and would like to offer you an opportunity." "Go on," I replied. The man continued that they had come across the World's Largest Horseshoe Crab in a CNN news story and wanted to make it a part of the movie, or at least some t-shirts of the crab. He had acquired my email and agreed to send all the information for me to examine when I returned to the hotel later in the day.

At the time, I was pastoring a church in the small town of Blanchester, Ohio where we had been given a rather large gift in 2006 following my speaking engagement at the soon-to-be-finished Creation Museum in Northern Kentucky. Two museum associates took me to the parking lot and said that they had been led to offer a massive horseshoe crab to our church. It was spread across a parking lot in nine huge pieces. I didn't know what to say. I mean, really, what do you do with a giant crustacean? They went on to explain that it had come from the Columbus Center Maritime Museum in Baltimore. The facility was so far in debt before it opened that they had to have a "fire sale" to acquire the funds to pay off their debt. Thus, Ken Ham, the man behind the Creation Museum and Answers in Genesis ministry, bought the crab and shipped it to Northern Kentucky.

That was in 1999 and we are now in October of 2005 during my speaking engagement. I went back to speak with my church about the unusual gift that we had been offered. We threw up our hands and said, "Why not?" The museum took care of the expense, into the thousands of dollars, to ship the giant model of a massive sea critter some hour and fifteen minutes up the road to Blanchester, Ohio. Five semi trucks and a crane lined the highway, and one by one the huge fiberglass and foam pieces were lifted off the flatbed trucks on a cold and drizzly day in February 2006. Let me make this clear, it was a sixty-eight feet long and twenty-four feet wide crab with a twelve-foot domed ceiling inside. After it was delivered to our church, we decided to use it for a creation-type attraction in a small rock and flower garden. This structure quickly became a tourist attraction that gained international fame.

Over the next three months, we gathered sometimes up to twelve men, along with a skid loader, to maneuver the pieces into place using a real horseshoe crab shell as a guide. Since the crab had been setting in a parking lot for six years, it was out of shape and didn't fit to its original design. It was going to take time for it to reform itself. During this time we had learned that a former sci-fi designer was responsible for designing and constructing the giant crustacean. Over the next three years the crab had gained national notoriety as one of the top roadside attractions in America, even attracting Evel Knievel's former body guard, now stunt jumper Gene Sullivan, to make a leap over the crab through a huge firewall in front of thousands on the site and millions on cable television.

Now, let's return to our day in the Smoky Mountains and with our conversation with the man from the film. He said to me, "I think this would be a great opportunity for you to reach out to teenagers and others like you never have been able to before." Reaching out! That was an opportunity that I would consider listening to. They wanted to use four of our t-shirts that we sold as fund raisers to feature in a quilt for the main character Bella. The quilt was to be given to her by her mother, in the film, as a graduation gift. I called the man, who was one of the production directors for the movie, and said I will have the shirts in the mail in two days and count us in. It was at that point in Pigeon Forge, Tennessee, that for some reason a thought came to me. All I can attribute it to was that God was whispering to me about a future opportunity to reach out to young people. I was game for that, and I realized that I was being led on another adventure to add to the growing list.

Then, if throwing a giant crab in a movie wasn't odd enough to get this day going, the feeling of the Ark of the Covenant came again. Yes, the Ark that carried the Ten Commandments! The Ark that Moses, David and Solomon had so reverently celebrated and protected. What was I supposed to do? I mean, I'm now setting in a hotel room in Gatlinburg, Tennessee with this need to research this incredible piece of history. Thoughts of *Raiders of the Lost Ark* and movie blockbuster *The Ten Commandments* consumed me. Why did I have this urge, this strong desire? Was it just a feeling, or was it a calling? I had accepted an opportunity to reach out with t-shirts in the *Eclipse*, but God cushioned it with an overwhelming urge that He had yet to reveal.

Paintings and Trumpets

Here's where things really became interesting. Later that day we visited a local art gallery in Pigeon Forge featuring a local painter who was very well known for his incredible work. His name was Spencer Williams, and his work in the window was beautiful. I walked in and found him sitting at a canvas next to the large plate glass window working on his next masterpiece. We began talking about his work and he described his portrayal of a Roman guard in a local passion play. I told him about our church and about our ministry with a great opportunity in the area we live to reach out to so many in need. Mr. Williams continued to paint as others entered the store to admire his work. His assistant wandered over and said, "Let me show you something I think you'll really like. This is why you really came in here today." Puzzled, I followed her around the corner where she stood in front of a beautiful painting. Gazing up, she said, "This is one that really reaches out to me." I stood mutely at this large painting on the wall. A beautifully painted face of Christ was in the upper left hand corner with a beautiful bold Lion of Judah lying on the ground with a crown and a large scroll next to him. Then, plain as day, prominently taking center stage, was the Ark of the Covenant. My first thought was, "Okay, coincidence." It wasn't an hour before when this overwhelming urge had come over me once again to begin research on an amazing historical artifact that consumes a great deal of the Scriptures and was now somehow beginning to consume my life. And now, a man I had only just met and his assistant take me to view a painting featuring the Ark. I scoffed it off as coincidence and moseyed my way out of the shop to visit a friend named Larry Stinson, who owns the Gold N Memories Jewelry shop next door. Larry, a master jeweler, and I had connected a couple years earlier on an amazing time to swap storied blessings and our friendship snowballed from that point on. After exchanging a few ideas we hugged and I went on my way to find my wife, Sherri, gazing through the areas many shops.

We left Pigeon Forge and ventured back to Gatlinburg, Tennessee, just a few miles from the busyness of its neighboring city. We made a left turn onto highway 321 and traveled to the Arts and Crafts district of the Great Smoky Mountains. Here you find pottery shops, painters, and a variety of incredible talents that attest to the ability of the artisans. As we stopped at one conglomerate of gift shops, my wife wandered off into some of the shops while I had more of a purpose to my shopping detail. I made a couple stops with my boys into a magic shop and then a concrete statue yard before going ahead and visiting a shop hoping to pickup something for my wife since she was preoccupied with crafts and pottery shops. As I looked around the owner approached and asked if she could help me find anything.

I responded kindly that I was just looking around. It was then that she said, "I might have something you're interested in over here." Not wincing, I turned and followed her to a rack of a few uniquely handmade necklaces. She selected one necklace in particular and said, "This is the one you're looking for." Of course, in my infinite wisdom, I smugly thought that to be a quick "sell this to the tourist line" and said, "No, I'm looking for something for my wife and I don't know if she would like that." From nowhere came the words that shocked me as she replied, "In the Bible in the book of Numbers, chapter 10, I believe, is the story of when God called on Moses to make two trumpets out of silver to call the people to His tabernacle. This is a handmade necklace of those trumpets of Moses." My eyes grew wide as I was thinking, "Here we go again." We are standing in the mountains of Tennessee and I have a woman whom I have never met before telling me of trumpets of Moses to call people to the tabernacle that housed the Ark of the Covenant. I began to wonder if these trumpets were actually calling me. I wasn't sure but I quickly said, "Thank you but I'll keep looking." She replied without haste, "No, I mean this is for you. Please take it. It's my gift to you." Bewildered and baffled, she placed the necklace over my head and around my neck. I stood there momentarily dazed and uttered a mild "thank you" and turned for the door with the necklace of the trumpets of Moses dangling around my neck and many thoughts racing through my mind. I turned back to her in the doorway; she nodded and told me to enjoy my time in the Smoky Mountains.

I thought that was surely enough for the day, but it didn't stop there. Two hours later, as the night sky began to shine and the downtown lights of Gatlinburg were brightly illuminating the streets, we decided to walk "the strip" and do a little shopping. The humidity had subsided and the temperatures were much more bearable which made for a very pleasant evening. Austin and Skyler were on their own adventures in Gatlinburg as my wife and I walked the sidewalks people watching and gazing at store displays. And then, very loudly, the theme song of *Raiders of the Lost Ark* began to belt out of one of the mini golf locations inside a building. I turned quickly and the man working the Adventure Golf attraction put his hand on my shoulder and said, "I don't have the Ark, but would you like to play some golf?" "What?" I exclaimed! He turned his sights to other passersby in his attempt to entice other customers, but my heart was pounding. The song and the comment left me a little shaken, but it was just another confirmation that God had something in the works that He was preparing me for. I quickly grabbed my wife's hand and moved away as fast as I could. We continued to move on and enjoy the beautiful evening together. Sherri wanted to know why I picked up the pace and scurried away so quickly. I shook it off as though I didn't notice anything and the topic was quickly forgotten. The next day it was time to leave the Smoky Mountains and begin my research. What was God trying to tell me? Where would this journey lead me? I knew it wasn't going to be an easy study, but I also knew it would be more than just another adventure. I knew this one may not be what I think it might. Something had changed for me during this visit to the Smokies; little did I know what God had in store.

Chapter 3

GOD ARRANGED MEETING

THE URGE OF THE ADVENTURE

After returning from the Smoky Mountains, I began to seriously research the Ark of the Covenant. It was funny, though, because I thought I knew all I needed to know about this ancient artifact. I had studied the Ark at the School of the Scriptures Bible College and was under the impression that I knew all I really needed to know. How wrong I was! I started where all searches should start, in the Holy Scriptures: Genesis, Exodus, Deuteronomy, Numbers, 2nd Chronicles; it seemed like the more I searched the Scriptures the more I found myself almost visualizing what Moses, the Israelites, Joshua, David, Solomon and Hezekiah were going through on each line of their journeys, battles, and prayers. But, I knew there was more to my search that I didn't understand, at least not yet. I studied about Moses' desire to serve the Lord and the frustrations he went through with the Israelites and their constant disobedience. Then I went to Joshua and his faithfulness to follow along with the commands of the Lord who dwelled between the cherubim on the Mercy Seat of the Ark. King David was an example on how to both treat and not treat the Ark of the Covenant and the consequences that must be paid when God's instructions are not followed. I studied the Israelites "Ark worship" instead of their faithfulness in just trusting God. Then I moved onto Solomon and his struggles and then to Hezekiah's prayer at the foot of the Ark. I extensively looked into the all important Mercy Seat of God upon the top of the Ark and Cherubim angels that kept guard over that Seat. This sparked another desire for study on the many conspiracy theories of the disappearance of the Ark. The theories were so numerous that I started to lose count of them. I became enthralled with the overall power that the Ark possessed and how incredible it must have been to witness its strength. It brought down the mighty walls of Jericho and parted the River Jordan. God's presence through the Ark slew armies and no one could touch it, let alone look at it except for the high priest in the Holy of Holies. I continued to research days on end to find out every last little bit of information I could on the Ark of the Covenant. It was all so important to my study, while I still was trying to figure out what God had in store for me.

This continued from August until early December 2009. Our church had planned an outing to the Creation Museum's spectacular Christmas display called the Bethlehem Nights in Northern Kentucky. After the outing, we stopped off at a pancake restaurant for a late night bite to eat where another conversation began. "Jim, why don't we start a study on some of the true biblical sites in the Bible that you are always talking about, like Noah's Ark and stuff?" one of those on the trip suggested to me. "Since you love this historical stuff, why not share some of it with us?" "Sure, why not? I'll start working on it Monday" I responded. I went home and began researching DVDs on some of the most reliable true sites of the Bible that have been discovered in recent years. The tomb of Jesus was one of high priority on my list. Then, Noah's Ark and Mt. Sinai, from programs viewed earlier on the History and Discovery Channels, as well as others. I located a variety of videos and one in particular was *Mountain of Fire* that featured a man known as the modern day Indiana Jones, Bob Cornuke. So I called the offices of the BASE Institute to order the video, and Cornuke himself answered the phone. He said he was leaving on a documentary shoot in Malta to continue his discovery of the anchors from what he, and many others believe to be the lost shipwreck of the Apostle Paul. Yes indeed! He had located the lost shipwreck from Acts 27 that left Paul and a full crew and prisoners swimming for the beach. Cornuke located all of the anchors from the ship and was getting ready to film the story. Bob said he would be happy to send us the DVDs and the conversation ended.

My search for biblical discovery DVDs continued, as did my research in to the Ark of the Covenant. In January 2010 I prayed about men in our church to help me build a replica of the Ark for an upcoming Sunday morning message I was planning to deliver. For some strange reason I felt the replica would serve to demonstrate my message. In my way of thinking, I thought the Lord had something for me to preach in order to reach some of the people in our congregation. The men I chose to build the Ark started to assemble it on a cold January evening. I didn't let them in on what we were doing before hand, but I simply gave them each instruction and diagram on what they needed to build. When we arrived at the church, the one I had asked to build the main box and lid had already done his work and left it sitting on a table in a small museum area that we had constructed. The others began to arrive. When I brought out the drawing of my plans, each of the men took a step back, gulped, and said, "We are building what?" My friend Terry Erwin built the box and lid, and I let him on the project ahead of time so he would have it ready for us. The others, all dear friends of mine (Raymond Rhinehimer, John Shoemaker, Tim Moore, Brad Kiphart and George Watkins) all had a specific specialty that they needed to complete for the project. Over the next week and a half we worked nearly every weekday evening, hiding the ark each night to keep it out of sight of anyone else in the church. Finally, on January 25th, 2010 the replica was completed as I put the last coat of gold flake paint over it and prepared it for Sunday's service. It was beautiful. I didn't show the men the completed project because I wanted it to be a surprise to them as well. When they last saw it, the ark was still just a box with trim around it. I took it from there and added the angels, poles and paint to give it the final look. It was amazing, authentic in nearly all respects.

The next morning was another typical day. I got up early and headed out for my daily routine, but I found out a couple of hours later that this was to become anything but a normal

Tuesday. My phone rang, and it was the man from the BASE Institute, explorer Bob Cornuke. He had returned from his journey to Malta and seemed very tired as he began to speak to me. But what he had to say to me was shocking and somewhat prophetic. Bob said, "Jim, during my trip to Malta I couldn't get your name off my mind. I am planning an expedition on the trail of the lost Ark of the Covenant and I feel like you're supposed to be with us." My first reaction was "WHAT?" Cornuke went on to say "I don't know what it is, but for some reason you are supposed to be on this journey with us." I didn't know what to say. I was taken back and surprised. Listen, this would be a dream come true, but it was a little odd to me that he would ask this with all that had happened up to this point, even the completion of the Ark, which had nothing to do with this phone call, just a day earlier. I told him that I would have to pray about it and that's what I did. For the next twenty-four hours I prayed and talked with my wife about the trip. She was just as amazed as I was, but not overly surprised. She had just said three days earlier that she felt our greatest adventures were still to come. I felt like donning a brown fedora, pulling out my western style whip hanging in my shed and throwing on the DVD of *Raiders of the Lost Ark*. I love those adventure films and that's all that was running through my head. With my exploration past, and the compulsion to study the Ark, I thought I knew where this journey was about to go. I picked up the phone and called my friend Terry Erwin and told him about this call. He quickly said that he was shocked to a point, but really wasn't surprised and had a feeling that God had more of a plan than just building a replica Ark; Terry remembered my story about the incident in Gatlinburg.

The next day Cornuke's wife Terry called me and said that after talking to Bob, she felt my wife Sherri needed to be with me on this trip. After hanging up, my phone rang again and it was another Terry. This time it was Terry Erwin, the only one who really knew this whole story from Gatlinburg to the building of our replica Ark. He said, "I've been thinking a lot about this guy asking you to go on this Ark expedition and I really feel that if you go, that Sherri should be with you." I paused, swallowed and finally said, "That's really strange, Terry, because just five minutes ago Bob Cornuke's wife, who is also named Terry, called and said that she felt Sherri should go as well." He concluded immediately, "Then it sounds like you have your answer." Later that day Sherri and I sat down and I relayed about the phone conversations to her. She was one hundred percent with me, a surprise considering she knew nothing about what we were getting ourselves into. Ironically, this call came the week we finished the ark replica, two days before my birthday, and five days before my message to our church on the Ark of the Covenant. I began to realize very quickly that our lives were about to change and head a direction that I had never imagined. Although somewhat hesitant, I was excited about this very quick turn of events. This was becoming very evident that we were about to enter into a new and thrilling chapter in our lives fulfilling a life-long desire to explore the world in search of adventure.

Finally, the Sunday arrived when I was to give my message on the study of the Ark. I had a dramatic scene set up as if it were devised for a movie set. After our worship music, and during the offering, I disappeared into an office behind the altar and quickly changed into some Indiana Jones looking clothing, ducked out the back door, and snuck around the side of the building. I had forgotten that the snow hadn't been shoveled on that side, and I was jumping through drifts up to my waste. The lights grew dim in the auditorium and an eerie music started to play. I came up the aisle and made my way to a pedestal near the stage. On the pedestal was a very small replica of the Ark in which I did the dramatic switch for it with a bag of sand. I made my way up the aisle and doors to the auditorium burst open. Instead of a large rolling boulder barreling toward me, my mother in her motorized wheel chair blasted through the door as I appeared to struggle to get away on my way to the stage, barely making an escape. Then I was attacked by two treasure seekers who were quickly defeated by my whip that one of our members, Zeke, planted in the audience. It was quite the scene and brought a great deal of excitement and laughter to start off the service. As I spoke about the Ark, I told the congregation the Ark looked just like this little replica that I had in my hand. Then I stepped back and said, "No, it looked more like this one." The doors burst open, and again as music played, four of the builders of the replica were dressed in ancient costumes and carried the Ark upon their shoulders with another leading them down the aisle. Cell phone cameras came out of nowhere as everyone wanted to capture the moment and a memory of this incredible full size gleaming gold box. The message was full of intrigue, surrounded by the Holy Spirit which led a young lady at the end of service to the altar and gave her life to Christ. Then before everyone left, I announced that Sherri and I had accepted the opportunity to go on this journey with

Mr. Cornuke and everyone erupted in applause anticipating this incredible opportunity.

That night the men of our church gathered together and decided that the church would find some way to help us finance a portion of the trip. This decision launched a series of speaking engagements and fund raisers as we prepared for our journey with Bob Cornuke and the rest of the Ark team. It was such a blessing to see the efforts of so many. It wasn't just an excitement for Sherri and me, but a building of excitement for our church. Everyone wanted to be a part of the expedition by helping to make it all happen. In my own mind, I was excited to see all of this, but couldn't help but wonder what else was in store for me. Somehow I knew that it wouldn't stop with just being involved in an expedition. I knew that God had more in mind when He started us on this journey some six months earlier. One lady said, "I'm not surprised, your life is an adventure every day." And honestly, it truly has been, but this one may be the setting for the rest of our lives. Time will tell.

Chapter 4

THE BETHLEHEM CALLING

It's almost a whimsical thought that a young man and his wife had to make such a long, tiresome journey to the village of his heritage in order to be included in a census for taxation. Today, we tend to run from the taxes, but at this time, that's exactly what young Joseph and Mary did as they traveled to Bethlehem.

"And it came to pass in those days, that there went out a decree from Caesar Augustus, that all the world should be taxed. (And this taxing was first made when Cyrenius was governor of Syria.) And all went to be taxed, every one into his own city. And Joseph also went up from Galilee, out of the city of Nazareth, into Judaea, unto the city of David, which is called Bethlehem; (because he was of the house

and lineage of David:) To be taxed with Mary his espoused wife, being great with child." Luke 2:1-5

A young couple, who made that commitment, knowing that Mary was not only "great with child," but Great with The Child, began this long journey that was sure to be full of hardships and criticism. This miraculous journey was not one that the couple really could predict the outcome, but they knew that it was something that had to be fulfilled. Ridicule was sure to follow them and trials were sure to plague them during this long trail they were about to travel. Either way, they knew what they had to do and were on their way to fulfill both the decree of Caesar and the prophecy of the birth of the Savior. This wasn't just going to be a trip to Bethlehem, but a life-long struggle that would begin with a short journey. Their minds must have been in a whirlwind during the trip. Wondering "why them" and "what was the next step?" As we know, the greatest journey for them was yet to come.

From today's perspective, to travel from Nazareth to Bethlehem would only take a couple of hours to complete the one hundred-mile trip in a car. Remember, we're talking two about thousand years ago and it was a mandatory request by Caesar Augustus that all were to return to their city of lineage for the calling of the census. This is a time, although many still travel this ancient way today in this region of the world, when they would have most likely made the journey on donkey, camel or horseback and even with that, only one of them could ride while the other walked. With an internet search or if you've visited this particular region of Israel, you will note a couple different types of terrain in this land today, which would have been more rugged during the time of Mary and Joseph. You either battled the desert sands or you dealt with rocky paths and mountains. Either way it wasn't easy by any means with rough trails, a scorching sun by day, and much cooler temperatures left them shivering in the night. A two-hour journey for us today would have easily been at least a ten day trip for a young man, his pregnant spouse and their belongings.

Most indications show that it is likely that the couple made their pass through the River Jordan Valley. While starting in Nazareth, approximately 1,200 feet above sea level, this trek would have descended quickly and most likely would have taken them across the river a couple of times before arriving in Jericho. They possibly would have passed by Waddy Kelt, where King David would have herded his father's sheep. This area, surrounded by canyons, is where many believe

David wrote Psalm 23 because of the deep valleys located there. From here, Mary and Joseph would have mostly likely ventured from the lowlands to the high mountains of Mount Olivet and either into or around Jerusalem *(ref. 12)*.

Another recent archeological find in this area uncovers a church that contains a rock believed to be where Mary rested with Joseph on their way to Bethlehem. As in many of the traditional sites in Israel, Jordan, Egypt and other parts of this region of the world, they are marked by traditional churches commemorating what is believed to have been the locations that either Mary or Jesus rested upon, touched, or it is noted as a site where some historical event occurred. There is no real way to tell, but the likelihood is that something was recorded to have taken place there at one time in history to document the claim. And even though scholars want tangible proof, many times the tradition passed down through the ages may be the only proof we may ever have and our discernment will have to lead us to make the qualitative decision of whether to believe it or not. Finally, Mary and Joseph would have arrived in the city of Bethlehem to complete the first leg of a life-long journey for God.

Shortly after their arrival in Bethlehem, Mary went into labor and gave birth to the King of Kings in a "so called" stable. We must look at this a little closer in order to gain a true perspective on the truth of their accommodations. Because the Greek word for "inn," in this case, is actually "*kataluma (ref. 4)*", which is more understood as a guest room, it's important that we understand what this actually would have been during this ancient time period. Since this journey was due to a decree sent out among the entire Roman Empire, the couple would have, more than likely, attempted to stay with relatives. But with the slower journey for them due to Mary's pregnancy, they would have probably arrived much later than the others in the family and were turned away from the guest room and sent to the lower end of the home where the animals most likely stayed at night. Considering that many homes were built into and above caves, the family would often bring the animals into the lower end of the home during the cooler nights. Mary soon went into labor and gave a miraculous birth to Jesus, the Savior of the world, in a humble setting in a far-away town.

Around the time of the birth of Christ, trouble was brewing in nearby Jerusalem. King Herod knew of the birth of a king but was not aware that it was The King. He probably didn't understand it anyhow with his pagan beliefs, but he was more concerned of the threat to his overall power and his throne. That was obvious from his next decree. With the inability to find this king who had been prophesized, the jealous Herod, after he realized that the wise men weren't coming back to give him the information he requested sent out an order to murder all male children from the ages of two and under.

"And being warned of God in a dream that they should not return to Herod, they departed into their own country another way" Matthew 2:12

"Then Herod, when he saw that he was mocked of the wise men, was exceeding wroth, and sent forth, and slew all the children that were in Bethlehem, and in all the coasts thereof, from two years old and under, according to the time which he had diligently inquired of the wise men." Matthew 2:16

It was near this time that Joseph had a visit from a Holy being;

"And when they were departed, behold, the angel of the Lord appeareth to Joseph in a dream, saying, Arise, and take the young child and his mother, and flee into Egypt, and be thou

there until I bring thee word: for Herod will seek the young child to destroy him. When he arose, he took the young child and his mother by night, and departed into Egypt:" Matthew 2:13-14

This was without a doubt a frightening call upon Joseph and Mary, to say the least. Remember, this was a couple who had just traveled from Nazareth to Bethlehem upon the order of Caesar Augustus. We know that this wasn't immediately after the birth of Christ because we see this when the Scripture tells us that the wise men came to a house and visited with a young child. No longer was Jesus a baby:

"And when they were come into the house, they saw the young child with Mary his mother, and fell down, and worshipped him: and when they had opened their treasures, they presented unto him gifts; gold, and frankincense, and myrrh" Matthew 2:11

So, from what we can gather from the Scripture is that Jesus was near the age of two when they began their next journey. As we see in Matthew 2:14, Joseph was instructed by the angel to take Jesus and Mary and flee from this area, completely beyond their comfort zone, into a much unchartered territory for them. This was a beginning to a miraculous and untold adventure that possibly prepared Jesus for all that was to come.

Scripture brings Joseph once again face to face with an angel sent from heaven:

"And when they were departed, behold, the angel of the Lord appeareth to Joseph in a dream, saying, Arise, and take the young child and his mother, and flee into Egypt, and be thou there until I bring thee word: for Herod will seek the young child to destroy him. When he arose, he took the young child and his mother by night, and departed into Egypt" Matthew 2:13-14

Herod's soldiers were on their way to slay the male children two and under and the angel warns Joseph to pack up this family and begin an exodus from the land they know and love. Rather than remaining hidden in their homeland, the Holy family would go to a land notorious for their worship of idols before massive structures erected to receive their worship. One can only imagine the fear and confusion that gripped Joseph and Mary as they crossed over to this mysterious land of Egypt with Jesus.

Without any doubt, though, Jesus knew exactly what was to come. He was not only escaping the slaughtering blades of the Roman soldiers, but He was on His way to meet with God the Father in another far-away land. During this miraculous journey, the prophecy that God's Son would enter into this land and idols would fall to Him was only the start of events. When you think about it, with their personal possessions wrapped up in a cloth, including the gold, frankincense and myrrh from the wise men, this family was about to venture into the unknown. They couldn't gas up the SUV or even water up a camel. All they had was possibly a donkey and their own company to count on. Isn't that the trademark of a family? They had each other and that's what really matters, not to forget that they also had the Lord of Lords in their presence. Really, when you think about it, what more do we really need for our own unknown journeys but Jesus, right?

Chapter 5

A NILE JOURNEY

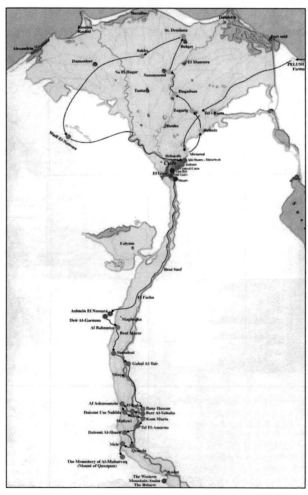

Jesus: In Ethiopia

No longer were Joseph and Mary a family from Nazareth, but now they were considered fugitives on the run from the evil king Herod and his fearless army. Although the Scriptures are silent about this portion of the life of Jesus as a boy, there are plenty of oral traditions and written documents to follow a very well defined path of the Holy family in the ancient land of Egypt. There are many credible historical documents that follow this path, traditional stories passed down through time, and many websites to journey through that now cover this topic. In our studies, we will take you through a historic path that was more a journey to follow the trail to somewhere already planned out, rather than to escape from the situation that awaits them if they don't heed to the call to move on. By now, I hope I have your attention to move further. Again John 21:25 states:

"And there are also many other things which Jesus did, the which, if they should be written every one, I suppose that even the world itself could not contain the books that should be written. Amen."

John tells us that there was much more that took place in the life of Jesus that wasn't included in the Scriptures. That's obvious since we have nearly thirty years that are untold in His life since the birth of Christ until He began His ministry on this earth. Or, at least we thought. And you will also find it ironic that it is John that makes these comments for what you are about to read in this book.

The Bible gives us what we need to become a Christian in faith, to turn our lifestyle to become more Christ-like, and to learn what is needed for nearly every topic that life has to offer. The books that are not in Scriptures were not needed for the application of righteousness in our lives and may not have been inspired by God to include. With that said, this exodus journey was more for the preparation of the ministry of Jesus to offer His followers until His eventual return. This adventure was not necessarily intended for Mary and Joseph, but for Jesus to prepare Himself for the introduction of His deity to the world. It's an amazing thought that nearly 2,000 years since Christ was born we are just now realizing that there was so much more to this King of Kings and His relationship with His Holy Father. We would be given glimpses of this many times in the Scriptures such as in the Garden of Gethsemane in Matthew 26:39;

"*And he went a little further, and fell on his face, and prayed, saying, O my Father, if it be possible, let this cup pass from me: nevertheless not as I will, but as thou wilt.*".

And then we would see it several times while Jesus was on the cross in Matthew 27, in John 17 and 19, Mark 15 and Luke 23:34;

"*Then said Jesus, Father, forgive them; for they know not what they do*" and in verse 42, "*Father, into thy hands I commend my spirit: and having said thus, he gave up the ghost.*"

We also see the reference to the deity of God the Father resting in Christ throughout the gospels, as well. It will also show us that the discussions Jesus had with God the Father in the gospels were not based on a short term relationship; rather it was the climax to the education that Jesus was taught earlier in His life.

The Beginning of the Journey:

Egypt was a different world for the Holy family. Carrying in their hearts the One and only God, Yahweh, was going to be a challenge for them in this new world of pagan beliefs.

The vast land of Egypt and its mighty empire bowed down to gods such as Baal, Astarte, Amun, Mut and Khonsu. These gods were worshipped in various locations throughout the empire of Egypt including the area known as Luxor, the Valley of the Kings. This is evident from the number of temples that can be seen in this once mighty city still today. The one thing that the Holy family had going for them, other than God in the flesh was traveling with them, is that the cultural acceptance of Egypt would be a welcoming relief as they made their trek through this country. Although there are many mentioned noted possible locations that the Holy family may have passed through, we will only focus on the ones that have a documented significance to their path and journey.

Today the land of Egypt is mostly visited for their great pyramids and statues to the pharaohs of the ancient past. However, a keen interest has emerged in the last few years to document the trail of Jesus traveling throughout this land. This has brought a different relevance to the interest in Egypt, one that is really well-defined. Each area where the Holy family passed, you will find churches or some type of commemorative memorial to identify and recognize their appearance in the village.

From all indications it appears as though Jesus, Joseph, Mary, and a previously unknown figure on this retreat (one that we will examine later) made their entry into Egypt from the north east region in the modern day location of Rafah. Rafah can be distinguished because of a lone ancient sycamore tree that is said to have been dated at the time the Holy family visited while passing through. It is important to note that many of the first few locations that the Holy family passed through were brief visits because of the threat of Herod's soldiers still on the search for the feared King.

After passing through Rafah, they moved onto the city of El-Arish for another brief respite. This is a city nestled on the banks of the Mediterranean Sea and the reference to their entering and exiting El-Arish is barely mentioned in any ancient writings, but we know they would have had to pass through here to move further on.

Although there were a couple other locations they may have possibly passed through, at this point in the Sinai region, their last location was in Pelusium, or Tell El-Farama. This is a place that was once a busy port-city, but now is a village of ancient ruins including numerous Roman churches. Again, this would have been a short stay for them as they moved on to escape the approaching soldiers of King Herod.

Their journey continued as they moved on through to one of the branches of the Nile River in the city of Tell El-Basta, where they probably passed through before resuming their flight through Egypt.

Al-Mahma, the bathing place as it is known, was their next stop. In Al-Mahma ancient history tells us that Mary bathed Jesus in this town. In a newly restored church, built in the 12th century, you will find the place where this event was to have taken place. Al-Mahma was not the most pleasing place to the Holy family though, as we will find in many of their stops along this path. It was un-safe and Mary and Joseph knew they had to protect their Son. They quickly gathered up their things and headed to the north to a friendlier environment in Bilbeis *(ref. 18)*

Footprint of the King of Kings

From here the family began to meet up with a branch of the Nile River and took a ferry boat to Sebennytos. For some reason they continued to move north, arriving in the town of Burullus in the Valley of Sysbann, where, again, they were not welcomed visitors here. The citizens of these villages didn't want the wrath of the Roman soldiers in their towns because of the possibility of having their homes ransacked if the Romans would have wanted to search further. So, they continued to move on.

This time they went to Al-Matlaah and then to Sakha, located in the western Delta, where a miraculous event would occur in this trek of Jesus through Egypt. In Sakha, during a time when Mary had Jesus in her arms, she became tired and sat Him on a rock. What significance would Mary placing her Son on a rock be, you may ask? The importance of this visit is that Jesus' footprint stayed on that rock and is still there today. This rock, preserved with sacred oils, had been hidden away for centuries in the fear that the relic would be a center of significance and target to be stolen. Sometime in the 1990's the "footprint of Jesus" was located and once again made available for viewing. The church that stands on this spot today still has the slab of rock that contains the footprint of a young child embedded in it. Is it the footprint of Jesus Christ as a young boy? Could an authentic incredible artifact have survived for nearly two thousand years? There's only One who knows the answer to that today. For us, our journey continues. Following the path of the Holy family is not an easy one, but as they traveled southward, it all matches up to a divine pattern of the miracles presenting the deity of Jesus, even as a young boy *(ref. 5)*.

Prophetic Desert Encounter and Fresh Springs For The Family

The Holy family ventured on, crossing the Nile into the western desert as they headed for the village named Scetis. Today it is known as Wadi Al-Natrun. On this path, in the desert, a prophetic encounter would take place, as laid out in the 1st Infancy Gospels (Chapter 8:1-8), and in Coptic writings as well. This was a migrating ground for thieves. Knowing this, Joseph and Mary made the choice to attempt to pass through this area during the night. During this quiet attempt to move through, they came upon two thieves asleep along the road and attempted to not awaken them. The thieves, Titus and Dumachus, were awakened and startled by the unexpected passers-by. Dumachus attempted to do what he did best and tried to rob the Holy family. Titus, realizing the company they were in, pleaded to Dumachus to let them pass without altercation. Dumachus continued, and again Titus, fearful of what might be put upon them from God Himself, pleaded with him by offering him gifts to let them pass. Mary recognized the kindness of Titus and said, *"The Lord God will receive thee to his right hand, and grant thee pardon of thy sins."* After the encounter, Jesus turned to His mother and said, *"When thirty years are expired, O mother, the Jews will crucify me at Jerusalem; And these two thieves shall be with me at the same time upon the cross, Titus on my right hand, and Dumanchus on my left, and from that time Titus shall go before me into paradise (ref. 6)."* This foretelling of the two men on the cross with Christ was a prophetic beginning that would come to involve many more foreseen predictions to come during their flight through this strange, but beautiful country.

After a short visit in Wadi Al-Natrun, they continued on to where Cairo is located today. They moved to the east bank of the Nile River to the city of Heliopolis. The name Heliopolis is taken from the Greek name for the Pharaonic city of On in the ancient times. Located today just outside of Cairo, it is called Ein Shams, when translated means "the eye of the sun." From here they traveled to Matarieh where a very unique event was recorded to have taken place. As the Holy family took shelter under a large sycamore tree, Mary sat next to the tree, and from out of the ground a spring of fresh water started flowing. If you visit Matarieh today, you will still see what they claim to be as "Mary's Tree," an ancient and decaying sycamore tree said to be the tree where she sought shelter with Jesus *(ref. 18)*.

Statues Fall and an Impression in Stone

The trek for the Holy family continued to the south with stops in Al-Zeitoun, Al-Zweila and eventually to Babylon or today known as Old Cairo. Local history states that when Jesus entered the town, the idol statues fell to the ground as prophesized in the Old Testament:

"Behold, the LORD rideth upon a swift cloud, and shall come into Egypt: and the idols of Egypt shall be moved at his presence, and the heart of Egypt shall melt in the midst of it" Isaiah 19:1.

This upset the people in such a way that the governor of the territory attempted to find them and have Jesus killed. Writings show that Jesus, Joseph and Mary hid in a cave which today carries some unique and distinctive marks to memorialize their visit. Today this cave is below the church of Abu Serga (or Saint Sergius). After their escape, the Holy family continued onto Maadi, to the south of Old Cairo, and then took a ferry boat across the river to Memphis, which was once the capital of Egypt. They boarded a boat from here to the town of Ashnein Al-Nassara where you will find a monastery on this site today where a deep well is located that allegedly supplied the Holy family with water. Four days later they journeyed onto a site known today as "the house of Jesus" or Abai Issous (present day known as Sandafa). From here they stopped in Samalout and then to the east bank of the Nile River to Gabal Al-Tair, known as "the mountain of birds." The family took up shelter in this spot in a cave where a church now rests above. As the historic writings show, while in this place a huge rock began to fall on the family, but Jesus held up His hand to stop its landing. Legend has it that this rock contained the embedded handprint of Jesus. Even though similar to the footprint in Sakha, this relic has been lost to the ages, for now *(ref. 7)*.

The Copper Idol Falls

Continuing on, just south of Gabal Al-Tair, there is an acacia tree whose leafy branches are said to have swept the ground and turned up as the Holy family passed by. The locals in that area call it the "worshipper" because it is said that this tree bowed to Jesus as He passed by during their journey along the trail. Continuing south the family arrived in Bir Al-Sahaba, where they are to have crossed the Nile River once again to further their trek southward. Eventually they came to Hermopolis Magna where a very kind man helped shelter the family. Taking them in, he risked a great deal of trouble because of his kindness. The reason behind the trouble resulted from the huge copper statue that fell when they entered the town. The people believed that evil spirits were in that idol and when it collapsed and broke that the spirits were let loose. The priests were angered by this and the family was forced to move out of Hermopolis Magna quickly. The kind man that sheltered them, on the other hand, was in a great deal of trouble for showing compassion to Jesus and His family *(ref. 18)*.

Death for His Faith and Palms that Bow

The Holy family moved on into the town of Al-Ashmounein, which brings us to an unfortunate end for one man proclaiming the deity of Jesus as He entered the village. A young man named Wadamun, who was from the village of Armant, received the spirit of God wrapped up in a boy's body as He entered the town and made it all known by declaring the deity of Jesus as they passed through. Because of his act of worship, Wadamun was put to death by the priests in the village. Although we see many martyrs later in the New Testament, this very well could have been one of the first to sacrifice his life by proclaiming Jesus as the Messiah.

Again, the Holy family would leave this hostile area and continue their venture to the south arriving in the Dairout Umm Nakhla, "the mother of the palms." As we saw earlier in Gabal Al-Tair, when Jesus passed through the town, the many date palms that lined the streets were believed to have bowed to Jesus as He passed by. In this area there is a very peculiar set of palm trees that again seem to be bowing in their structure, unlike many others surrounding them along the pathway still today.

Abu Haneis was the Holy family's next stop as they crossed back over the Nile River again to the east bank. Here they stopped for a period to quench their dire thirst from the belting sun and crystallized desert sand. At Abu Haneis, the well that quenched their thirst was named Sahaba. This is translated "cloud," thus named to the credit of Mary "moving like a swift cloud" in search of water for her Son. The reception here was much more encouraging than many of the previous stops along their trek. With that welcome they decided to take some time here and rest on a hill that is still named today Kom Maria. Nearby, a church still stands commemorating this visit by the Holy family, the Church of the Holy Virgin *(ref. 18)*.

Hostility and Welcome

Eventually their trek continued southward, crossing the Nile once again, and arriving in the town of Daitout Al-Sharif, or Philes in ancient times. With word spreading who this child claimed to be, hostility grew and the priests became anxious and incited the people to drive the Holy family from their village. Again, as they journeyed onward another hostile encounter took place in Al-Qusseya where they were driven out because their godly claim cast fear upon their pagan worship lifestyle. It seems ironic that even when Jesus was a boy, they were driving Him out of their lives. Unfortunately, still today, the world seems to continue to drive Him out of their lives.

Finally, the Holy family arrived in Meir where they were received with open arms. After a short stay, they traveled to al Muharraq, or in ancient times Mount Qussqam. In Egypt, they actually stayed at this location for the longest period of time, for six months and ten days as noted in historical writings. We must also note that because Muharraq is not the farthest region to the south that it is claimed the Holy Family visited, we must assume that the time period they spent here may have been split between a visit to and from their final destination to the south. It is reported that after they passed through Muharraq, and were welcomed, that Mary wanted to return to this city. The word Muharraq means "burnt" because the grasses in the area were set on fire to enable the farmers remove vegetation and to plant crops *(ref. 5)*. In this town, you will find what is reported to be the first church built in Egypt over the cave where the Holy family took shelter during their stay. The Coptic Christians never consecrated this church because they claim that it didn't have to be blessed because Jesus consecrated it by His presence there. This was a favorite location of Mary, and the family considered this home for a period of time.

From al Muharraq they continued southward to Dayr el-Maymun where the villagers believe that Jesus blessed the village as He passed through there. Continuing south the Holy Family ventured into Wadi Al-Rayan and took up shelter in a series of caves. These caves were known to house hermits and were also heavily used as a perfect hiding place in later years for Christians dodging the Roman persecutions. They then ventured by Sawada and to Asena, Dayr al-Barsha, Dayr Abu Hinnis, Buq, Assiut, and finally their farthest point in Egypt in Dayr Rifa. This Egyptian village contains an incredible cave church that was built later to recognize the Holy Family's visit there *(ref. 20)*. Many in Ethiopia believe that this was the final stop for Jesus before continuing to Ethiopia for the prophetic meeting at Lake Tana.

The Journey Home

Although nearly every bit of evidence points to the Holy family living in Muharraq, I have to believe that since they enjoyed their stay there, that after returning from the future prophetic meeting between The Father and The Son in Ethiopia, they returned to the safety of Muharraq to wait for the calling to return home. Egyptian writings claim that the Holy family stayed in Murharraq until Joseph received the calling by the angel to come to Israel, I can't help but question why Joseph would take his family so far south, through hostile surroundings in many cases, to then be told that it was time to come home. There had to be a purpose for such a long journey. Herod's soldiers, in no way, could have followed them that far to the south. Jesus would have known that, and God wouldn't have put them through the struggles of traveling if He didn't have another purpose for them. This is also one of the clues that led me to question why they would travel so far to the south knowing that there would be a long return home. What was the purpose of this journey southward?

Best-selling author, Paul Perry, shared with me many of the highlights of his exploration to Egypt to study the flight of the Holy family. But, when asked if this was the last stop, the word assumption emerges. Paul's book *Jesus in Egypt* is, without a doubt, the best source for extensive research on the topic of the trek of the Holy family. To his credit, his research was specifically about the Holy family in Egypt. That's why nearly every source discussing the flight of the Holy family stops in or around Muharraq before heading back to Israel.

That really fueled my inquisitive mind enhanced my questions of "Why?" Why such a long journey and what were they still running from? I couldn't get past the thought that God would send His Son and His earthly family on such a treacherous and long journey. It didn't add up to me. But, tradition controls so much of our thought process. Little did I know my questions were about to be answered.

The Coptic writings tell us that the long journey for the Holy family home was ready to begin. After leaving Ethiopia they would have traveled back to Assiut and then ventured back to Muharraq. It is there the writings tell that Joseph received the calling of the angel for them to return home.

"But when Herod was dead, behold, an angel of the Lord appeareth in a dream to Joseph in Egypt, Saying, Arise, and take the young child and his mother, and go into the land of Israel: for they are dead which sought the young child's life. And he arose, and took the young child and his mother, and came into the land of Israel." (Matthew 2:19-21)

It is believed they sailed the Nile River to Memphis and docked at Al-Badrashein. They would have passed through Maadi and Heliopolis and then through the desert until they arrived in Nazareth where Joseph felt it better to go, rather than return to Jerusalem because they feared Archelaus, son of Herod, the new ruler. As they came out of Egypt, though, it again fulfilled the prophecy of Hosea *(ref. 18)*;

"...then I loved him, and called my son out of Egypt." (Hosea 11:1)

Many believe that when the Holy Family left Egypt they may have joined up with one of the many bands of merchants that traveled the desert. The reasoning behind this is that their fear of Herod's son seeking Jesus and to use safety of the caravan. These caravans also traveled a much easier and safer route. The road home through Judea and Samaria would have been extremely rough. The caravans traveled a road known as Via Maris, or the "Way of the Sea." This road connects Egypt and Damascus and enabled the merchants to freely travel to all areas in Middle East *(ref. 7)*. There is much conflict on the time that the Holy family spent out of Israel during this journey. The Copts claim that it took around three years to complete this trek. The Muslim belief is that it was around seven years for the flight. My speculation is that the Muslim belief is more accurate based on the travel distance to Murharraq and then the ultimate goal of Ethiopia. Regardless of time or distance, it was a journey of Heavenly calling.

Chapter 6

TO THE LAND BEYOND

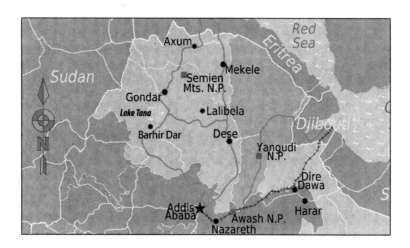

The first question that the critics will pose is that the Scriptures say nothing of Jesus going any farther than Egypt during His family's retreat from Herod's soldiers. That's true, but we must also realize that those same Scriptures give us no insight into where Jesus went in Egypt or what He did while He was there. The Bible only tells us that they went into Egypt and nothing more. One man quizzed me on the ability of Jesus as boy. "Do you believe Jesus could perform miracles and received His later godly power as a boy?" At that moment, I became more defensive in my response than I should have. I was more shocked than anything that someone would think that Jesus was all of a sudden "awarded" His godly ability when He matured. Or maybe God would snap His fingers and Jesus would magically be given the title of Son of God. When that question arose I was still somewhat unclear about some of the things I had been shown on my journey and needed more time to research further. Yet, spontaneously I blurted, "Yes, I believe Jesus Christ was God in the flesh from the moment He was conceived."

The Land of Ethiopia

Let's take a look at the ancient land of Ethiopia and its relation to the Bible with a look into both the desolate and beautiful features of this land. First, Ethiopia is a place like I have never experienced before. Shrouded in tradition, mystery, love, desolation, and beauty, Ethiopia brings the wonders of the ancient times with some of the luxuries their world has to offer. Although the capital city is Addis Ababa, where you see incredible poverty contrasted by the wealth of the palace of the country's president, and the embassies of many of the world's powerful countries; however, we will not be focusing on this part of the Ethiopia. Addis is located a great deal south of where I believe the ancient land of Ethiopia rests today. The lands we hear about in the Scriptures are primarily in the northern mountainous regions, meeting up with Sudan. It's the area that connects the feeding point of the Nile River to Egypt. It's a land spread with beauty, poverty, mystery, death and disease. That description might be a broad range, but, that's exactly what this land is made up of. The people are some of the most loving people I have ever experienced in my travels, while they are, in many cases, suffering from HIV, tuberculosis, mal-nutrition, alcoholism, and a variety of other horrible conditions. But one thing is true; they have an abiding faith for the most part.

Let's take a look at that faith. To look at this correctly we have to go back to Noah and his family. After the great flood waters had receded into their shores and banks, Noah's sons and their wives were told by God to disperse and replenish the earth.

"And you, be ye fruitful, and multiply; bring forth abundantly in the earth, and multiply therein. And God spake unto Noah, and to his sons with him, saying, And I, behold, I establish my covenant with you, and with your seed after you;" Genesis 9:7-9

They packed up their belongings and began their journeys to the outreach parts of the earth. Shem, Noah's oldest son, and the one that Jesus allegedly descended from, went out with his family to what is now Israel and the area of Jerusalem. Japeth took his family to Asia Minor and the modern day Europe and began their new life in that region. Then, Noah's youngest son Ham moved into Canaan, which is Saudi Arabia, Jordan, Egypt, and the Eastern Africa area. One of Ham's descendants settled the area of Ethiopia, historically one of the world's oldest and most ancient countries.

Looking at this region you will find the oldest region of the Aksumite Kingdom. This was a mighty and powerful kingdom established around 1000 BC in northern Ethiopia and the eventually spread downward to the central portion of the country. It was during this time that King Solomon, who reigned from 970-930 BC, received a visit from the Queen of Sheba according to 1 Kings 10:1-13. Enamored by her beauty, the King surrendered to her charms, and the Queen conceived a son whose name was Menelik. Many theories surround Menelik and his involvement with the Ark of the Covenant. One theory says that Menelik came back to visit his father in Jerusalem, and Solomon showered his returning son with favor and riches during the visit. The people of the kingdom also generously welcomed Menelik. This undue attention greatly concerned the priests, and they went to Solomon and warned him that Menelik would have to go because of the danger of the people's devotion to him. Solomon agreed but stipulated that their first born must go with him to Ethiopia. In this order, the son of the high priest removed the Ark and many other items and stole them away. During their travels to the Axumite Kingdom, Menelik was informed of the items that accompanied them.

Another theory speculates that Menelik stole the Ark of the Covenant from Solomon and took it to his mother as a gift. Neither tradition carries any merit whatsoever because we can see that the holy Ark remained in the temple of Solomon long after his death and during the time of Hezekiah (2 Kings 19:15). Nevertheless, Solomon had appointed Menelik the King of Ethiopia after he returned to Jerusalem to receive a blessing from his father. This is the most commonly accepted

Jesus: In Ethiopia

belief of how Ethiopia received Judaism, which soon spread throughout the country.

The Falashians, or the black Jews, began to practice this form of Judaism, but they were secluded with only the Pentateuch, the first five books of Moses, and centered their belief solely on these writings. Axum (Aksum), Ethiopia became a great city and a mighty kingdom which had a direct tie to the trading routes of the Roman Empire with the connection of the Red Sea at the eastern banks of Ethiopian's northern coast. Legend has it that two Syrian boys shipwrecked there and became slaves. After gaining favor with Emperor Ella Amida, these two boys began to teach Christianity to the king. After the king's death, his son Ezanus (or King Ezana) took over the mighty throne around 300 AD and the Christian faith began to spread. Ezana appointed one of the boys, Frumentius who was now an older man, as the bishop of Ethiopia. In approximately 305 AD as the bishop, Frumentius began to spread the Christian belief throughout the land of Ethiopia. He even led King Ezana to the Christian faith. King Ezana also plays another role of importance to Axum history, which we will review more closely in chapters to come *(ref. 14)*.

Axum, Ethiopia

Now that we have looked at the arrival of Christianity in Ethiopia, let's look at this country today. The northern region of Ethiopia is made up of a variety of villages and cities with spiritual and historical significance. Keep in mind; the cities are not the thriving metropolises we are accustomed to in the United States by any stretch of the imagination. For the most part these have seen better days since their mighty years of powerful kingdoms centuries before. One of the best examples of this is the city of Axum, where many of the mighty kings reigned, including The Queen of Sheba, King Ezana (Ezanus), King Kaleb, and others. One of my most amazing discoveries during my visit is that much of the ancient kingdom is still buried under centuries of sand, dirt, and rocks. Nearly every year the Axumite farmers will stumble across another structure, palace, or remnant of the original mighty Axum Empire, yet no regular archeological digs are consistent there.

Today, in many cases, this is a poverty ridden and diseased cursed part of the world. Whenever I visit there to continue my research, I am literally in tears over the suffering of these people. But, without any doubt, they have become survivors. They have a love and a care for each other like I have never seen before. They also reach out that hand to you. From the minute I stepped off the plane the first time and journeyed into this ancient land while dodging camels and donkey carts and watching as thousands shrouded in their robes walked through the dusty deserts and over the rocky mountain sides to do their daily routines, it didn't take long to realize that these people have a faith and a love lost in the materialized countries of this world.

Home of the Ark?

The village of Axum, Ethiopia is without a doubt one of the most mystifying places on this earth. In these highlands, desert location, it houses some of history's greatest treasures. Unlike the well preserved artifacts and historical sites that we encounter in the United States, in Ethiopia they are, in many cases, recognized but still unprotected and withering away. When you arrive in Axum, you quickly feel like you either dropped back in time 2,000 years or they just said "action" in an Indiana Jones movie. Other than small taxies and work trucks, the inhabitants of this ancient land walk, ride on donkey-pulled carts, or by camel. Dusty, rocky roads and open markets are the common place. White robed adorned people are common place and the buildings and structures stand in an apparent time warp. There is a real biblical-like feeling to this ancient place that totally defies description.

However, this remotely nestled Ethiopian town does make a most dramatic assertion to its credit. Axum is the only place in the world that claims to house the Holy Ark of the Covenant. Housed in a daubed chapel, Saint Mary of Zion Church, it is the resting place for the legendary Ark. The building is surrounded by a high iron fence with pointed tips and is heavily guarded. The interesting thing about Axum is that they don't publicize the fact that they have the Ark and they really don't care if you come to their city or not, as a visitor or tourist. They simply claim that God has allowed them to be the keepers of the Ark until the time comes that it will be called upon again.

The Guardian of the Ark, chosen for his purity and direct line to Solomon, gives his life to stay within the fence and guard the Ark until he dies. Once chosen, this man never will leave the grounds of this church again and agrees to completely remove himself from his family and friends. Even though it's an honor to be chosen as the Guardian (only thirty thus far in history) it is also somewhat sad to see his apparent loneliness. But, that is from a westerner's point of view. The honor that he holds as the only human allowed to view the Ark of the Covenant, may outweigh that he has given up everything in life to serve God in this way. When you truly grasp the situation from that point of view, maybe that's the ultimate responsibility of worship in the first place, to give up everything we have in life to serve God as Christ asked us to do in the Scriptures.

In one particular corner of Axum, you will see a collection of massive towering stelae (or obelisks) to commemorate each of the mighty kings or successful battles of the Axumite Empire. On this site you will also find the tombs buried underneath of some of the great kings of the Axumite Empire. During the occupation of Mussolini in 1937 during World War II in Ethiopia, the dictator wanted to claim the Ark of the Covenant. The story has it that when the Ark was lowered and whisked away into the hundreds of miles of underground tunnels under the church of the Ark, that Mussolini dismantled the second largest of the stelae and took it away on trucks. For many years it stood at the Piazza di Porta Capenemin in Rome, near the Arch of Constantine. In 2005, the 160 ton obelisk was finally returned to its rightful place in Axum where they have taken great care to re-erect the structure in its original resting place.

A few hundred yards away is the legendary bathing pool of the Queen of Sheba. Even though the water is extremely dirty and contaminated with disease, it has become a constant water supply, a place to wash clothes and to bathe. It also serves as a key spiritual site for the ceremony of Timkat, recognizing the baptism of Christ. Next to the bathing pool is a rocky road that moves up the mountain to the excavated, but not completely revealed ruins of the palace of the Queen of Sheba. Along that same trail are the palace remains and tombs of other great kings such as the mighty King Kaleb and King Gebre Meskel. Just a few yards past their tombs is a breathtaking view of the Adwa Mountain range. Near the great stelaes are other tombs including the mighty King Ezana, who is credited with bringing the Christian faith to Ethiopia.

In their possession they also claim to have many artifacts that came from Solomon's temple along with the Ark of the Covenant. If this were true, I knew that I had to see it for myself. We were given the special permission to go into the treasury because of Bob Cornuke's favor with the high priest, Narub. Within their treasury of the Ark, they house the solid gold crowns of the kings, the robes that adorned these rulers, their reigning staffs, along with many artifacts sharing the history of this proud country during their many great years of power and favorability.

While looking at the crowns, as the rest of the group went farther into the building to view the reported covering that was over the Ark when it arrived in Axum, I started to sweat profusely. My hands began to tingle and numb, and I started to feel somewhat sick. Since I was alone, I felt it best that I should go down to my knees, which I did while turning around as I dropped to the floor. Then I saw them. It took me back to Gatlinburg when the woman put the necklace around my neck, where it still rested. Everyone else had passed these by. Before me were what appeared to be two large silver trumpets, each nearly five feet in length. I called to one of our translators for an explanation, who in turn called the curator of the treasury who said, through Misgana's translation, "These are the trumpets of Moses from what is said to be from Bible's book of Numbers. These trumpets came to Axum with the Ark of the Covenant from Solomon's temple. These are the hammered silver trumpets of Moses." Even though the numbing stopped, the sweating began to increase as I started to think, "Am I kneeling in front of the two actual trumpets that Moses had in his hand? Am I really looking at these while the Holy Ark is just fifty yards from me in the building next door?" It was obvious that these trumpets were ancient and pounded. Each of them had rings around them to possibly signify the twelve tribes of Israel. I asked if I could hold one. The curator understood and simply smiled, shook his head, and said, "No, no." I quickly grabbed my wife who shared my emotion, and she began to tear up as well. This became a theme to our team shortly after leaving the building. "When the trumpets call, will we arise and go?" This awareness escalated my excitement, and my anticipation of future adventure gave me an adrenaline rush that will never leave. To this day, I have not lost the passion for this adventure, this calling of the trumpets of Moses.

Why would God choose such a forsaken place to house His precious Ark of the Covenant and His throne, the Mercy Seat? It's a place riddled with poverty and illness, yet abounding in love. It's a place that time has forgotten yet one of the most amazing visions I have ever seen. It's not a typical vacation site, but in its own way it's a place of beauty not only for its landscape of desert mountain views, beautiful eagles and hawks soaring through the skies, and ancient artifacts and sites, but, because of the people that inhabit this land. These are a people who have been trodden down and forgotten, but carry a proud and unselfish beauty with them through their lives. Maybe this would be the perfect place to hide the Ark, until the time that it will be called upon again.

Lalibela, Ethiopia

Then we make our way to Lalibela. This small city is one of the most incredible places on this earth. Nowhere else will you be able to see the amazing sites found in this ancient village in the upper mountainous regions of Ethiopia. History has it that the rightful heir to the throne at Roha was a young boy named Gebre Mesqel Lalibela. At birth, a swarm of bees surrounded him and his mother named him Lalibela, which means "bees recognize his sovereignty." This also was a sign to his mother that he would one day become the emperor of Ethiopia. Lalibela was in conflict with his uncle, Tatadim, and his own brother, King Kedus Harbe. He was also nearly poisoned to death by a half sister during this inner family battle. During this time, Lalibela went into exile in Jerusalem for several years.

During his time there, he gained great knowledge and felt that he was given a vision to bring the Holy land back to his home and to retake his rightful throne. Many historians believe that since Lalibela's brother was still alive when he returned to his homeland that he took the throne by force. So the self-proclaimed rightful heir to the throne was now the King of Ethiopia and began to transfer Roha into a New Jerusalem while making it the capital of Ethiopia. He brought with him olive trees from Israel and renamed the hillside the Mount of Olives. He also changed the name of the small river that flows through the town to be called the River Jordan.

Considering that the entire city is on top of a mountain of pink granite, this is what makes Roha (later named Lalibela to honor their king) different from any other place on earth. In this mountain top, poverty driven, community you will find the most amazing carvings ever to be completed. Somehow King Lalibela in just over two decades carved these eleven rock-hewn churches from the mountainside. One of them is the famous Church of St. George, or Bete Giyorgis, which is shaped like an Ethiopian cross and can only be entered by following a tunnel and choosing the right direction of a cave to pass through. Then, after making the correct decision, you will finally come out to the open area of the church. Many believe that this church was constructed to house the Ark of the Covenant which never arrived there.

The largest of these structures is the Bet Medhane Alem church which is said to be the home of the Lalibela cross. It is also said to be the largest monolithic church in the world with its massive pillars and many rooms within. This church also claims to have the secrets of how these churches were constructed in such a short period of time. Lalibela claimed it was with the help of angels while others have many other theories. The secret to their construction is said to be beyond a tall pillar inside the church in the area known as the Holy of Holies. This section is heavily guarded and no one is allowed beyond the shrouded pillar to investigate. This particular church is also believed to have a model, or copy, of the original church of the Ark, Saint Mary of Zion Church, in Axum. Another one of the amazing churches worth mentioning is Bete Golgotha which is believed to be the final resting place of King Lalibela, himself. No matter who built them and how they did it, the churches of Lalibela are an important and credible wonder of this world.

The journey to Lalibela is not an easy one, but one that was well worth trekking around the world to see. This city is one of the holiest cities in Ethiopia, second only to Axum because of its belief to be the home of the Ark of the Covenant. This amazing place left me awe struck as I stood on the mountainside next to ancient olive trees and looked around at beautiful pink granite monuments that in many ways lay in tribute to a former king and his inspired vision to recreate the Holy land in his home town. Regardless of who helped this king carve out these beautiful and mesmerizing structures, one look at their sharp edges and near perfect construction will show that God's hand had to be in this masterpiece. Some say it was the Ethiopians who worked together to create these beautiful structures. Others claim that the Knights Templar traveled to Ethiopia in search of the Ark of the Covenant and were the architects of churches that mesmerize the modern world today. While others, including Lalibela claimed that this was the work of the Ethiopians and of angels that helped complete the structures in just twenty-three years. The churches were so perfectly made that each one has been constructed with a natural water flow underneath the rock to keep the wells filled with water. How did Lalibela do it? That's a question that still baffles the architects, the engineers and the world centuries after their construction *(ref. 16)*.

Gondar, Ethiopia

This is truly an interesting place, known as the Camelot of Ethiopia. Gondar was founded by Emperor Fasilides around 1635 and feels in many ways like you're in the middle of a medieval novel. The claim to Camelot comes from their many castles built by former Emperors who reigned there. This became the capital city in Ethiopia and Fasilides built his palace there along with seven churches scattered throughout the city. No one really knows why Fasilides chose Gondar to build his empire, but his claim to this place brought life to not only the city but also the region that surrounds it. Five emperors who followed him also built their palaces there. One factor to understand about most of the kings that proceeded this time period was that many dwelled in temporary royal encampments and had no permanent palaces. There are exceptions to this rule, however, with the powerful empires that have since been uncovered in areas such as the Axumite Kingdom and others.

During a visit to Gondar, located north of Lake Tana, you will notice a distinct difference in this city. You will be quick to notice the castles all throughout the area, along with a great deal of Italian influence because of the Italian occupation during their conquest in 1936 and again as Italy's last stand in World War II against the English. You can still see ruins from the wars including one particular site that houses Mussolini's bunker atop one of the mountains.

The incredible castles and palaces are show-stoppers in this ancient city. It has been the basis for many battles and traditionally was divided into quarters for its diversity of residents. One of those residents includes Beta Israel, or more commonly known at the Falashian Jews. Although most of these Ethiopian Jews were air-lifted to Jerusalem during a series of evacuations in the 1990's, this is the site in which their origin was based, dating back to the time of Moses, with ties to the Queen of Sheba's son, Menelik. The Jews in this region are heavily persecuted and most have been killed for their claims and beliefs. It's believed their direct lines to the Levite Tribe may be the ones responsible for the returning of the Ark of the Covenant to Jerusalem upon Christ's return.

Gondar is a place that provides magnificent views of the rough African terrain while standing atop the mountains that surround it. Even though poverty and disease are seen throughout this location, as it is in all of Ethiopia, Gondar is truly a gem resting among the stony landscape. The castles offer up glorious days of knights riding into the kingdom while a sense of intrigue fills the air about some of the darker days that were sure to have passed in and out of the ages of time during the reigns of various emperors.

One of the show-stoppers in Gondar is the Debra Berhan Selassie Church which is more commonly known as The Light of the Trinity Church. It was built by Emperor Eyasu II in the 17th century and today is one of Ethiopia's most important churches. During a battle in 1888, the Mahdist Dervishes of the Sudan ransacked the city and burned down every church in the city with the exception of Debra Berhan Selassie. Legend has it that when the forces approached the church a swarm of bees flew down surrounding the church and thwarted the attempt to destroy the structure. It is also believed by the Ethiopians that during this swarming of the bees, the Archangel Michael stood with a flaming sword as if to guard the wooden gates leading to the church (*ref. 13*). Either way it still holds to an exciting story of the existence of this stone-walled and thatched roofed church with some of the country's most beautiful painted walls and ceilings depicting many biblical scenes, the trinity, and hundreds of angels covering from top to bottom.

Although the poverty can be seen at every turn, Gondar was still uniquely compelling for us during our travels. As a matter of fact our guide, Misgana, purchased a large portion of property atop one the highest mountains in Gondar. His plans are to build a resort there to attract visitors from around the world to bask in the intrigue, beauty and mystery that surround this medieval city.

Barhir Dar, Ethiopia

BarHir Dar, located on the southern banks of the mighty Lake Tana in Ethiopia, is unique for many reasons. Not only is it a popular tourist spot in Ethiopia, because of Lake Tana, but also the lake itself is the start of the Blue Nile River and the beautiful Blue Nile Falls. Barhir Dar is known for their palm-lined streets and friendly people. At the same time it is a heavily populated Muslim city which adds a different perspective and feel than most other locations we visited in Ethiopia. Palm trees can be seen throughout the city as well as jacaranda trees which are in full bloom in February and add a colorful touch to this flourishing city in Ethiopian standards. Despite the high volume of the Christian faith in Ethiopia, there exists a looming concern throughout this region over the last several years.

The Blue Nile Falls remains one of Ethiopia's biggest attractions. This natural wonder bursts out of Lake Tana and is known as 'the smoke of fire' falls which somewhat describes their beautiful spray hovering over the expanse of the lush tropical setting with an occasional rainbow splashing

Jesus: In Ethiopia

color into the mist. Lake Tana is another desirable site to see in this region of Ethiopia as it is nestled in the mountainous area of the country. When flying to this massive lake, you will notice the arrival well before you even see the lake because of the lush countryside of green that beckons you. Lake Tana is one of Ethiopia's largest lakes measuring over fifty-two miles long, from the north to the south, and over forty-one miles wide, east to west. It has an average depth, considering the dry season, of forty-nine feet deep with a plentiful stock of fish. The lake is also the starting point to one of the most famous rivers in the world, the great Nile River, which runs northward through Egypt. The Blue Nile River, which later meets with the White Nile to form the Nile River, begins in Lake Tana and is home to nearly three dozen islands, including Tana Kirkos. These islands contain many monasteries and are only inhabited by monks who appear to be somewhat guardians of the islands. Some of the islands contain lost scrolls, tombs of emperors, and ancient palace or temple remains.

Barhir Dar is the primary city docking against the lake and gained its popularity and growth from the masses that come to take in the beauty of this pristine body of water. But Barhir Dar held other purposes preceding World War

II. After the Italians took over Gondar in 1937, they moved south to Barhir Dar. The English Royal Air Force bombed Barhir Dar in 1940 causing the Italians to retreat back to Gondar. One interesting note is that throughout Ethiopia, ruins of destroyed buildings and vehicles can be seen dating back to their many conflicts, but also because of the remnants of the 2nd World War *(ref.15)*. In Barhir Dar you may have to look a little harder because of the recent tourist market in the city.

With nearly a quarter million people living in or near this city, it continues to be one of the biggest cities of growth in the country of Ethiopia. Then, of course, it is believed that at one point in time, around 410 BC, the Ark of the Covenant would have moved through this area, before the founding of the city, and rested well off the shores of the lake on Tana Kirkos Island. This island is restricted to visitors and one can only be welcomed there with special permission from the monks that guard it today. Emperor Selassie's (Ethiopian's King from 1930-1974) palace is located near Barhir Dar. He was responsible for building one of the large highway bridges over the Abay. Even though this city is unique and beautiful, it is one that is quickly losing any grip it has on Christianity.

The Finishing View

These are just a few of the sites of the northern mountain regions of Ethiopia, but probably the most common to the Holy nature of the country. So much turmoil has taken place in this region over the centuries. Both civil wars and world battles have both used Ethiopia as a ground of conflict throughout the ages. Great wealth turned to great poverty since the times of the early kings of this great nation. But one thing is for sure; the northern region of Ethiopia has been a hot spot of spiritual quests for adventurers throughout the years. This truly could be the location of many of the God's mysterious secrets from the housing of His throne and His law to an incredible meeting with His Son. It is an overwhelming feeling that this could be the place on this earth where we will hear those trumpets blast again, and the world will see a sign from these mountains that will be heard and seen from around the globe.

Chapter 7

BELIEFS and WRITINGS

Once Christ was resurrected from the grave, He began an intense series of teachings to His disciples of specific instructions of where to go, what to do, and how to accomplish all of these directives. First they had to believe on Him and then they had to wait until the arrival of the Holy Spirit to fill them with the third element of the godly Trinity. Previously Jesus would tell His followers that He will be sending another to dwell in them forever;

"And I will pray the Father, and he shall give you another Comforter, that he may abide with you for ever; Even the Spirit of truth;" (John 14:16, 17).

After Christ appeared to His disciples, He instructed them to go to Galilee, and He would meet them there for further guidance and to put aside all doubt that He was alive. After He gave them their final direction to *"Go ye into all the world, and preach the gospel to every creature,"* Mark 16:15, the disciples went on to do what most of the world struggles with. They now had to wait the arrival of the unknown, unseen Spirit that was going to guide them in their mission to reach out to the world with the good news of Jesus Christ.

That day was soon to come. With the biblical account approximately ten days after Christ ascended into Heaven, the believers were about to be hit with what Jesus referred to as the Comforter. As they gathered and waited, the moment finally came. The Day of Pentecost had arrived, and the Holy Spirit came down on the disciples and they were now on their way preaching the good news of Christ, performing miracles and showing signs and wonders to all they came in contact with. These followers, filled with the Holy Spirit, traveled throughout the towns, cities and countries, beyond their areas of comfort while avoiding the religious leaders of the day, who were not believers in Christ and persecuted the disciples because of their threat to their cause and the control of the people. That is exactly why Jesus told them to preach the gospel and that "*ye shall be witnesses to me both in Jerusalem, and in all Judea, and in Samaria, and unto the uttermost part of the earth*" Acts 1:8.

Mark and the Egyptians

We will begin to look into the beliefs of this particular religious region in Egypt since the Holy family ventured into the land of ancient statues and pyramids towering out of the desert sands. In this sector of the world is the Christian Coptic Orthodox Church located in Egypt. In order to help us learn the differences in this faith, we will break this Coptic Christian belief down from its beginnings.

The word Copt came initially from the word Aigyptos from the Greek language translation. This word was taken from the adaptation of one of the names for Memphis, "Hikaptah" which was the ancient capital of Egypt and was founded in around 3000 B.C. by Pharaoh Menes. Today the word Coptic is the definition of the Christians in Egypt. It also is used in as a description of some of the most amazing architecture and artistic paintings that came directly from the faithful artisans of these Christians throughout the centuries *(ref. 16)*.

The apostles traveled throughout the occupied world spreading the gospel of Christ across the countryside and into new lands. Mark was no exception to this rule. During the reign of the Emperor Nero, one of the Roman Empire's most notorious rulers, Mark brought the teaching of Jesus to the land of Egypt about ten to twenty years after the day of Pentecost. The Christian faith spread like a wild fire on a windy dry day throughout Egypt. Within fifty years after Mark began his ministry in Alexandria, historical accounts show newly found New Testament writings discovered in Bahnasa, Egypt and even fragments from John's gospel writings written and transcribed into the language devised by the Coptics *(ref. 10)*. Some of these Scriptures have been dated back to the first half of the second century. The Coptic Orthodox Christians of Egypt believe that their church was the intent of the prophecies found in Isaiah 19:19, "*In that day shall there be an altar to the LORD in the midst of the land of Egypt, and a pillar at the border thereof to the LORD.*"

The Strength of the Coptic Church

The Coptic Orthodox Church in Egypt has grown dramatically over the last nineteen hundred years. One of the most intriguing features of this church is their dedication to preserving their Scriptures, writings and authenticity to their oral traditions. Not until the late eighteenth and early nineteenth century did many of their oral traditional teachings finally become transcribed onto paper, papyri or animal hide scrolls. But that is not to say that the stories and the writings are flawed. Throughout time, priests would recite and teach the oral traditions to the congregations. Those who heard the teachings would in turn, relay the priest's message to others in the community, which was sometimes misquoted but quickly corrected by others who sat in the presentation given by the Holy men preaching the biblical Scriptures. Fortunately, this accuracy has been passed down through the ages and has been a key to the preservation of many of the writings describing the trek of the Holy family in Egypt.

This is also the same procedure behind such documents as the Dead Sea Scrolls and their accuracy. When the nine hundred and seventy two texts, known as the Dead Sea Scrolls, were discovered in a series of caves in Qumran, in what is now called the West Bank, between 1947 through 1956, it became the single most important find to authenticating the Holy Bible. These scrolls have been dated back to 100 BC and were spread over eleven different caves throughout the mountains. Previously, the oldest documents that were on record were dated at 1100 AD which makes the find of the Dead Sea Scrolls most significant to authenticating the translation we use today. Nearly every book of the Old Testament was found in these caves except for the book of Esther. One of the most amazing finds was the scroll of Isaiah. Miraculously, the comparison of the two documents that were transcribed nearly twelve hundred years apart, were ninety-five percent accurate in that comparison. The only differences were in their spelling or in grammatical issues, while realizing that the meaning was kept the same throughout. This is not only amazing in its own right, but also proof to the precision and care to making sure the transcribed Scriptures were accurately passed from generation to generation. This is the same process used by the Copts and by the Ethiopian Christians in their traditional scrolls and books in the Coptic Uncial and from the Ge'ez into Amharic today.

The Coptic Church today is a very strong defender of the faith. They have maintained that role since the first century because of their trust in God Himself sending His Son Jesus there when Herod ordered his soldiers to murder all the children two years of age and younger. The

Coptic Christians have played a large role in the building of Christianity throughout the ages. They have had a huge influence on Christian theology and beliefs over the centuries and have the ancient documents dating back to the days when the disciples spread throughout their lands *(ref. 17)*. They have been strong defenders of protecting the faith from the Gnostic heresies and are responsible for thousands upon thousands of biblical and theological studies throughout the centuries. Many of these studies have been used in archeological research because of their accuracies.

The Coptic Church translated the Holy Bible into the Coptic language in the second century and still today universities and museums around the globe safeguard these accurate biblical manuscripts. This church has been a mainstay in the historical preservation of the Bible and its contents. Their strong belief in both oral and written tradition has been a cross check for them and many others over the centuries. And, as archeology continues to uncover new artifacts and biblical sites the accuracy of the Coptic manuscripts will continue to play an intricate and essential role in verifying biblical truth.

The Coptic Cross

The Coptic Church has been a survivor throughout the ages in the ancient land of Egypt. They claim that their ability to survive takes them back to 68 AD when Mark was dragged by his feet through the streets of Alexandria by the Roman soldiers and ultimately killed. They staunchly believe they are to defend their faith as Mark did until his death. These groups of Christian believers have been persecuted throughout the centuries by nearly every ruler since that time including the torture, exile and even death of many of their priests and monks. During this mistreatment, the Copts formed, or created, a symbolic cross to represent their belief. In the original form and present day variations, the circle in the cross represents the everlasting love and eternal nature of God. This same symbolism is used today in wedding ceremonies with the representation of the circle of the ring and its representation of God's everlasting love and eternal Spirit *(ref. 10)*.

In most forms of the cross, you will find a halo to depict the deity of Christ and the overall cross to show the crucifixion of Jesus in Jerusalem. The Coptic cross is carried by nearly every monk and priest in Egypt, as well as in Ethiopia's Orthodox Church. One particular note about the Ethiopian crosses, which is the same in Egypt, is that there are many variations of the cross and many of which are designed to represent the village, or city, in which they serve in. Today these crosses are a popular collectible for tourists to gather the many different designs from the various villages.

In addition to the cross, other Coptic traditions came to light during this time of persecution. Roman Emperor Diocletian was a notorious persecutor of the Christian faith. It was his exiling of the Coptic Christians into the Sinai desert that brought about the Calendar of the Martyrs, which is still used today by farmers in Egypt to keep a course with the many agricultural seasons, and for the Uncial language, which is a virtually extinct and ancient language. The growth of the Coptic Christian faith in Egypt was strong, sweeping the majority of the country, until the twelfth century when the Muslims began to take priority. This conflict between the Muslims and the Christians would many times burst into violent battles.

Today, nearly ten million Coptic Christians exist in Egypt with thousands of churches still thriving throughout the country. Plus, there is another one and a half million Copts serving in branch churches in various countries across the globe *(ref. 10)*. The Egyptian Coptic Christians are prayer-warriors and continually pray for the uniting of all Christian Churches to work together for the cause of Christ. An interesting point is that we see this in the earliest writings in Alexandria, as you will note in a translation later on. They also spend a great deal of their prayer time asking for blessings on their country, the continued flow of the Nile River, their agricultural crops, their leaders and government, and their people. They are peaceful people and their prayers of peace are examples of the lifestyle that they live. The Ethiopian Christians also pray for the flow of water continuing from Lake Tana into the Blue Nile River which is the beginning of the Nile River. We could take a great lesson from their teachings because of the continual peace that they have preserved throughout time

The Ethiopian Orthodox Christians

Just as the Egyptian Christians are peaceful, there are none that rival the overall peace as that of the Ethiopian Orthodox Christians. From the moment I first stepped foot in the northern mountain regions of Ethiopia, it was evident that the people were loving, caring and very hospitable. That hospitality dates back centuries even though these people have endured civil war, disease, famine, and many other conflicts, including World War II. Through each adversity, the people of Ethiopia have become more peaceful in nature, but only up to a point. The turning point comes when their faith is threatened as a result of their protection of the Ark of the Covenant. Because of the intensity of their faith I thought it essential to research the Ethiopian Orthodox Church and learn where their foundations came from to better understand their culture.

Although much of the beginnings of the Ethiopian Church is lost or hidden, the true foundation comes somewhat directly from strong Hebrew heritage throughout the ages. One may ask, "How can its foundation be Hebrew, while claiming to be Christian?" If that was your first reaction, then you have a valid question. As described in this writing, the direct connection to the Levite tradition of the Old Testament tribe of Israel is strongly seen in Northern Ethiopia. Since many claim to be direct descendants of the Levitical line of the Hebrews, you will most commonly see these claims when it comes to direct dealings with the Ark of the Covenant or its Jewish background in general. The Falashian Jews, which are spread throughout this region, with the heavier concentration near Gondar, are believed to be of the ancient line to the same Levites involved in

Jesus: In Ethiopia

the transportation of the Ark, and possibly descendents of Emperor Menelik, the son of Sheba and Solomon.

Either way, the Hebrew heritage came together in history with the Eastern Christian foundation to bring Ethiopia its claim of faith today. Even though the western world churches have gone through many changes and ever-rising variations of denominations over the last two hundred years, the Ethiopian Church has changed very little over centuries of their faith. This goes primarily back to the beliefs of the Copts in Egypt and their strong conviction to keeping the faith through their strong traditions of preserving their Christianity while accepting their Jewish roots.

With their intertwined relativity to each other, both Hebrew heritage and Christ's teaching through the Apostles, you can see the Hebraic, characteristics of Hebrew language and culture, through their worship. Legend has it that when Queen Makeda, or better known as the Queen of Sheba, journeyed back from her encounter with King Solomon in Jerusalem and gave birth to their son, Menelik I. Before this journey she was heavily into the African region's belief and trust in numerous gods, very much like the beliefs that overcame Egypt. But, the legend has it that upon the birth of her son, she was overwhelmed with the faith of Solomon and eventually condemned the worship of familiar pagan gods, and claimed the full faith in the One true God. Sheba's story with Solomon can be found in 1 Kings 10:1-13.

Another theory is that when Menelik returned to Jerusalem to visit his father, he took the Ark of the Covenant into his possession and brought it back to Ethiopia accompanied by Azarias, the son of Zadok the high priest of Israel, along with all the first born of the house of Israel. While others believe that the most likely transition to the Christian faith into the Hebrew culture came after the introduction of the gospel by the disciples Mark and John to this land. Following that event, the full spread of Christianity is believed to have commenced when King Ezana gave his tutor Frumentius, a Syrian Christian appointed by his father when he was a boy to teach his son, as the head of the Ethiopian Church. At this point Ezana made Christianity widely known and elevated it as the official faith of the nation. This, along with their customary Hebrew teachings, became mixed and combined throughout the ages resulting in the overall strong faith in the Old Testament, and understanding of the New Testament, within the people there today.

Their strong Hebrew beliefs and claims, mixed with the teachings of the Orthodox Church tradition, have created an interesting and somewhat unusual worship over the ages. Their faith in Jesus to take away their sin is strong, but their trust in the Ark of the Covenant is somewhat overwhelming. During my time with these people, I witnessed women bowing in front of the Church of Saint Mary of Zion where it is believed the Ark of the Covenant rests today in prayer in order to conceive a child. If they should become pregnant after this prayer they believe it to be a blessing from God who dwells in the Ark and will honor Him by giving their first born child to the church to be raised as a monk. These monks throughout time have been the keepers of the traditions and the protectors of the Ark of the Covenant. Their Hebrew claims to Solomon and the Levitical high priesthood are clear, but the mixture of claiming the Sainthood of Mary, the forgiveness of Christ and the worship to the Ark of the Covenant have become something in appearance as a worship of the Ark itself. Similar to Old Testament worship of trusting in the physical Ark over God's instruction from above, the Ethiopians have, over time, become a combination of many beliefs and structures.

Regardless, the people have developed an incredible passion for their faith and one that needs to be adopted by the western churches. Not a worship of objects, but their worship to God in prayer and their dedication of their lives to serve Him. It's quite possible that God has made them live this way in order to protect and serve Him and to serve as a role model for us. We see this truth in Isaiah, Ezekiel and even in Psalm 68:31; *"Ethiopia shall soon stretch out her hands unto God."*

Their lives center around their worship, unlike that of the western world where our lives center around ourselves and God is slipped in only when needed. Their belief that this nation has been given the privilege of guarding the Ark of the Covenant is not only centered in Axum, but throughout their country. Nearly every church throughout most of Ethiopia has a replica of the Ark, or Tabot, in their Holy of Holies and even the replica is regarded as an extremely sacred object. The word "tabot" is a Ge'ez word which means a replica or the Tablets of the Law, or sometimes referred to as a replica of the Ark of the Covenant. I have studied their faiths and seen the differences in Ethiopia, Egypt and other faiths as well. Without a doubt the Ethiopian faith has combined strong ancient Hebrew heritage with the teachings of the apostles, while conforming to the object their country has been entrusted to guard *(ref. 16)*.

Ancient Language of Ge'ez

The ancient language of Ge'ez has nearly become an extinct language to many throughout Ethiopia. Just as the Uncial Coptic is to the Egyptians and Latin is to the west, Ge'ez is the same to the Ethiopians *(ref. 8)*. Just as the other two languages mentioned above were used for a very long time as standard language, Ge'ez was the primary language of Ethiopia and it was also the base for the Semitic languages of Amharic, Tigrinya and Tigre that are used today in various areas of the country. Although Tigrinya and Tigre are used sparingly, Amharic is the most widely used and is the official language of Ethiopia and is spoken most often in the mid to upper mountainous regions of the country.

The ancient language of Ge'ez was made up of twenty-four symbols and based on the Sabean system of writing. This system included the writing in alternate lines in opposite directions. This was to say they moved from left to right and then going the opposite way on the next line, alternating back and forth. Before Ethiopia's conversion to Christianity in Axum in the fourth century, during the reign of King

Ezana, vowels were not in use in this language. Thus, following this conversion to possibly make the language more understandable to those who were newly able to read the Scriptures, the vowel system was added to the language *(ref. 8)*. The Greek Bible text was then translated over to Ge'ez, which was very similar in the organization of the Greek.

For the majority, the Ge'ez language lost most of its luster just before the tenth century, AD, and the divisions and evolutions of the other three became the spoken languages of the regions. Even today, Ge'ez can be found in many of the Ethiopian Bibles, and most definitely in the ancient scriptures found throughout their country which have been preserved in the archives of the thousands of churches. This same theory is true with the Latin language and its existence as the spoken language ceased between the eleventh to twelfth centuries.

Today, Ge'ez is primarily a language in a biblical sense that is spoken by the old priests of the land to keep the ancient translations accurate. In the following chapters, we will see how this language is not only somewhat lost, but how it has been lifted to a language spoken and understood by those in a sacred manor with the integrity of Ethiopian biblical church translations. During a closer look into this language, one interesting similarity that I noticed when trying to decipher a nearly thousand-year-old Ge'ez Bible was its closeness to the Egyptian hieroglyphic system, which has been referred to many times as the basis for its origin *(ref. 8)*. My time in the courtyard in front of the Church of Saint Mary of Zion in Axum, the alleged home of the Ark, brought me to this awareness. In this area considered to be holy ground are many thrones, columns, and tombstones. One of the stones that caught my attention was engraved with what appeared to be hieroglyphic writing. One of our guides, Sumara, told me that this was probably a priest or judge that had come from Egypt and that this was his "stone of his death." Although Ge'ez doesn't have the symbols and drawings of ancient hieroglyphics, it does present a story line with every line, circle, wave, and dot.

Ge'ez is a very hard language to trace its origin. It is commonly believed to have been a writing with its foundation in the Asian to Middle Eastern origin, rather than that of the traditional African languages. Even to the Ethiopians that we encountered in the northern mountain areas of the country, we found that Ge'ez is only read by the older priests and monks and has a symbolic, sacred and holy nature to its existence today. It's looked upon as a language of old and is kept in a sacred reverence and is spoken and translated by the holy men that oversee and keep the tradition of the Scripture close to their hearts. Understanding that Ge'ez is nearly an extinct language to most of the world, with Amharic as the base language today, it's almost a spiritual feeling knowing that this text could possibly be hidden away forever in the pages of scrolls of ancient scriptures and historical documents of a long lost Ethiopian culture. Or, just maybe only hidden until the appointed time God feels that more of the mysterious of the Bible are ready to be unveiled.

Chapter 8

THE BANKS OF A MIGHTY LAKE

Somehow I knew as I was standing on the cliff overlooking the village of Axum on our final night before embarking on the next leg of our journey, that our adventure had really only begun. As the sun was setting with a bronze finish spreading over the land, I reviewed the last few days in my mind and referred back to my journal. It was at this point that it really started to hit me. We had explored the tombs of ancient kings, in a rare opportunity met with the Guardian of the Ark, fell to my knees before the incredible hammered silver trumpets that the monks claim to be those of Moses, took part in the Timkat ceremony, and experienced a spiritual rush when we witnessed to our young guide and translator, Sesay. Through all of this remembrance, I couldn't help but think that something else was still to come. By this time the sun had dipped behind the mountains and a chill had come over this land. I slipped on my coat and was joined by Dave Kochis on the balcony; he soon saw the thought process that was obviously written on my face. He said, "It's really wonderful here, isn't it?" "It all is!" I replied. "But, Dave do you ever feel like God took you somewhere to remove you from everything you are comfortable with just so that He can shut you down long enough to show you something really big?" Dave exclaimed, "Yes, and I think that is why we are both here, but, for different reasons. I feel God is helping me from a personal stand-point, and I think He's got something else in store for you." That left me in even more wonder. Sherri came out to join us, and she quickly saw that I had tears in my eyes and had been in deep thought. She is always quick to see these things in me. Sherri can always read me whether I'm hurting, worried, fearful, angry or happy. Arm in arm we told Dave goodnight and made our way down the dark rock steps, past an old sycamore tree to our room for the night. Laying there with my eyes staring at the ceiling I couldn't wait to see what was next. I spent many nights in Ethiopia staring at the ceiling. The information overload could only be processed in the darkness and silence before retiring each night.

Morning arrived and we loaded up our van and worked our way down the mountain back through Axum to the one runway airport in the ancient village. Before we pulled into the airport, we couldn't help but notice the half blown-up remains of a bunker that was more than likely remains from the Italian battle against English in World War II or from one of their many civil wars. As we turned to our final roadway to the airport, a young armed guard dressed in blue camouflage stopped us and ordered us out of our van. He wanted passports to confirm our identities. Even though toting a machine gun and trying to give stern instructions, the soldier was more than willing to take a photo opportunity to display his gun with our group. I believe this example proves the love of these people and their passion for peace in their hearts.

After the short delay, we were again on our way. Misgana's friends came to help with the luggage and get it to the inspectors before loading on the plane. I heard someone say in broken English, "Who dis bag belong to?" I knew it had to be mine. So I headed back to the entrance, and the guard had me open up the bag to reveal its contents. I was a little worried considering that I had a very old Bible in there along with two swords measuring about two and a half feet long and around 200 years old that I obtained from an antiquities dealer in Axum. But he bypassed both of the swords and kept looking. Then it hit me, I had also made an unescorted return into a couple of tombs of King Kaleb and of King Ezana. In both cases I had located ancient stones in the walls of the tombs. All I could think of was that this guard had seen something that looked like jagged objects and became inquisitive with what he saw. I tried not to show any emotion, but I was worried that they may have seen something that I may not have been aware of. The Bible I wasn't as worried about because I had wrapped it in my underwear and stacked my personal Bible and other books of reference around it. I knew they wouldn't have anything to do with my underwear, so I felt pretty safe with that. And since the guard had passed by the swords, I was assured it was the stones he was searching for. But then he reached in and pulled up a round adventure camera in a clear tube designed for underwater use. I was carrying with me several cameras, three of which were able to be concealed. He asked what this was and I took it out of the casing and showed him. I think he was so intrigued by it that he completely lost the thought of dissecting it for further examination. He shook his head, smiled and waved me on. Sweat had really started to build up on my neck, and after he waived me on, I took a small rag from my back-pack, wiped the sweat away, and moved on. Without a doubt, I realized that I needed to rearrange things for future inspections.

Marye on the Mountain

We boarded a turbo prop plane and started on our journey to Gondar, located north of Lake Tana. As we cleared the clouds we were immediately on the runway, whisking side to side in frantic stop before running off the end of the pavement. A man on a camel rode by waving as we exited the plane and walked across the tarmac to the building. Misgana's men were there waiting for us and loaded our bags as we hopped on another van and made our way to Gondar. Remember, this city carries a great deal of its ancient roots, as the palaces, or medieval castles, still reign high today. Our purpose for visiting Gondar was not to take in the architecture, but to head to the top of one of the mountains in search of a woman and her village, who claims to be one of the most ancient areas of the remaining Falashian Jews.

As we scurried up the rocky dirt road I noticed a series of very large, four to five foot tall, mud statues of the Lion of Judah with a Star of David on their heads. It was like each little hut had one in their front. Bob Cornuke jokingly said,

"The first person to point out Marye (the Falashian woman we were searching for) wins a prize." I stepped out of the van and a feeling overwhelmed me and I turned and said, "There she is, Bob." He stared at me as if to say, "How did he know that." Although I can't answer that one, I knew this was the woman we had come to see. We learned of her family being killed trying to return to her from Israel and how the rest of the Falashians on that mountain had been killed, or disappeared, for their beliefs. We sat in a make-shift hut as Marye, through Misgana translating, shared her sad story as tears rolled down her face. Larry, one of the men on the expedition with us, took a cloth and reached over to wipe her tears. Seeing her pain while attempting to share the agony of her struggles, persecutions, and the loss of her family was beyond belief and almost made us all forget why we had come. But now it was time to continue what we came to do. Kathryn Pierce, a nurse who had been a part of the exploration team, pulled out swabs to get an inside cheek sample for a DNA test to be done back in the states. It was all to determine if Marye was truly one of the Falashian Jews and a possible direct line to legendary Levite Jewish tribe. Marye was timid at first, but allowed Kathryn to continue with the swabbing. Just when I thought it was time to go, my wife and I turned back. In her small hut home were Sherri, Misgana, Marye and I. I looked at Marye as I asked Misgana to tell her it was a privilege to be in her home. Misgana introduced me as a holy man from America and what happened next was nothing less than a miracle. The woman who only could speak in the ancient Amharic language leaned over, grabbed my shoulders and said to me, "I know Jesus." Dazed and confused, my wife and I stepped back and looked at each other, as if to say, "Did you hear that?" But, what was really confusing is that our guide, Misgana, never flinched; he had heard nothing. Sherri, still filming, stood with her mouth open and tears streaming down her face. Marye's tears were non-stop and at this point, I really didn't know what to do next. We turned and headed to the opening of her hut and made our way outside. In front of her home was a large stone fire stove which was also used as a kiln for firing her pottery. I noticed this as we entered and that it had recently been used. Marye had a little wood handmade shelf sitting right outside her door with small terracotta pots and miniature Lions of Judah with the Star of David on their heads, matching the much larger ones lining the dirt road coming up the mountain. With Sherri still filming, Marye reached over and grabbed one of her coffee pots, looked for my wife, and handed the pot to her as a gift. Nearly dropping the camera, she hugged Marye, again in tears. Then Marye turned to me and handed me a Lion of Judah. It was an incredible feeling to know we may have just witnessed God speaking through this woman. This was a woman who had such a visible sadness about her, but a calmness to her demeanor that you just don't see unless someone is filled with the Holy Spirit. Later study of our video footage would show that Marye spoke in her native tongue on the video, but when it came to our ears it was plainly in English. It was an incredible blessing to be with Marye and one that will always stay with me for the rest of my life.

Following our visit with Marye, we headed across the mountain to the highest point of the range to where Misgana had purchased land in order to build a future resort on the site. As a boy, Misgana had been taken away from the island where he was raised on Lake Tana and forced and beaten to serve as a child soldier. To look at him now, a successful Ethiopian guide can only make you have hope for anyone in any dire situation to succeed. Misgana has not given up from his early struggles and is now on the verge of a life-long dream. One quick glance and you will see that something was different about the mountain top that Misgana has given his life savings to develop. All around are the remains of military bunkers on his property. Misgana pointed out to us some of the features of his resort when I asked him what one particular, blown-apart remains of a building was. He explained that in World War II it was the bunker of Benito Mussolini. Or, at least, what was left of it. This was the highest mountain point and the Italians used it as a stronghold in their final battles against the English. By the looks of it, Mussolini's bunker didn't fare very well in their final battles. I think the English may have gotten the best of this particular bunker. The view was beautiful and it didn't take us long to see the excitement of Misgana for his investment and his future vision. It will be a state of the art resort with shopping, a theater, pools and much more. Believe me, unless you're staying at the Sheraton in Addis Ababa, you won't find anything like this anywhere in Ethiopia. We wished him well, anxious to return some time in the future to see the result of his many years of struggle and hard work.

Traveling to Barhir Dar

After our descent down the mountain, we began the long journey to the south. As we traveled through the sky-scraper mountain tops the story-book-like African huts could be seen all around. At one point we had a desert on the left and a jungle with monkeys screaming out on the right. One thing we found very quickly, you better have a good ear, because if you don't hear the horn of the rare oncoming vehicle you would quickly become road-kill. Since there are very few motorized vehicles on the roadways, you see sometimes people by the hundreds walking along the roads leading their herds, traveling on camel, donkey carts or carrying massive loads of goods on the shoulders, heads and backs. Many times, especially when passing through the villages, the driver would beep the horn, never leaving his path or slowing down, and if you were ever in the way, Lord help you! They do not stop for anyone or anything. I asked the driver at one point if he had ever hit anyone and his reply nonchalant was, in his Amharic accent, "Many." That was a little scary. I encourage you to keep this in mind if ever walking along the road in Africa. If you hear a horn, move quickly or you might become another notch on the belt of the wild drivers of Ethiopia.

As we traveled, we came to a massive stone tower powering out of the hillside like a New York City skyscraper and straight up several hundreds of feet. One of our translators said it was called the finger of God. This was in the area where the mountains protruded high and the lushness of the landscape greened up under it. The "finger of God" looked like a high rise building exploding from the trees and peaking into the Ethiopian skyline. It was near Lake Tana as we could see the sparkling of the water at certain points glistening on the other side of the mountains. It was so picturesque that it appeared to be an illusion. We made a stop at the "finger of God" for a bathroom break. I escorted Sherri about a hundred yards off the roadway to a large rock and stood watch for her. Then, rather than standing watch for me, she wandered back to the roadway and left me standing alone behind the rock. As I stood there out of nowhere from the desert a boy appeared and waved to me as he walked by. I don't know where he came from but the encounter didn't phase him anywhere like it startled me. When I returned to the group, I saw people swarming everywhere. It was like people were moving rocks and crawling up out of the ground. I don't know from where, but our vans were surrounded by children hoping for a sticker or piece of candy. There was not a hut or house to be seen for miles, but from the hillsides these kids appeared. It was an awesome sight; this precious memory still lingers with me. As we continued traveling, I started noticing caves in the mountain sides and realized that these kids probably called those indentions in rock home.

Finally after a half a day's drive we arrived in Barhir Dar, which rests on the southern bank of Lake Tana, one of Ethiopia's largest lakes. Without a doubt it is a rather unusual town. Each morning you will hear the chanting of prayers, almost rivaling each other, broadcasting over loud speakers from both the Christian churches and the Muslims. The people are gracious, but a very different feel than that of Axum. After arriving at our motel, Sherri and I followed our adventurous nature to check out the area. My wife and I headed out to the bustling town square. As the taxis frantically hovered around the circle square, it reminded me of a life size version of the old video game Frogger. It was hit or miss just to get from one side of the street to the other. Although many that we came in contact with carried the Ethiopian kindness that we had grown familiar with in other areas, it was not the majority in Barhir Dar. We were definitely an outsider, and we could feel it.

Bible Study Following The Ark to Ethiopia

After returning to the motel, our research team gathered under the canopy near the lobby for a Bible study. It was during this time that a reality began to set in. It was a reality that I began to understand that this mission wasn't

Jesus: In Ethiopia

necessarily about the Ark of the Covenant, but much more than that. Just like Dave had said to me on the cliff-side in Axum, we each had something God was trying to open up to us. He said that his was a personal reason for being on this expedition, and my reason was going to be a little different. As Bob laid out the Scripture from the Bible to the reference of why the Ark could very well be in Axum and how it spent some time near Barhir Dar, it started to lay on my mind and heart that all I had seen thus far was a pathway of bread crumbs leading me to this mysterious lake. It was a continuing thought, similar to a skipping record on an old scratchy record player, which kept jumping back to each step of this journey and right up to this night. Then it would all start again. I knew that the next day would bring a revelation that none of us were expecting. It was like the night couldn't end soon enough to get on with the adventure still to come. Bob laid out the Scriptures of the theory on the Ark of the Covenant being housed in Ethiopia as a holding place for its return to Jesus. He shared that according to Scriptures; it was definite that at the time of Hezekiah, the Ark remained in Jerusalem at that time:

"And Hezekiah prayed before the LORD, and said, O LORD God of Israel, which dwellest between the cherubims, thou art the God" 2 Kings 19:15

Then, more than likely disappearing around the time of the reign of the evil King Manasseh, Hezekiah's son, who defiled the temple and brought disgrace to Israel,

"And he did that which was evil in the sight of the LORD, after the abominations of the heathen, whom the LORD cast out before the children of Israel. For he built up again the high places which Hezekiah his father had destroyed; and he reared up altars for Baal...and worshipped all the host of heaven, and served them. And he built altars in the house of the LORD, of which the LORD said, In Jerusalem will I put my name. And he built altars for all the host of heaven in the two courts of the house of the LORD" 2 Kings 21:2-7, 16

Bob explained that at the same time a temple modeled after Solomon's had been built on Elephantine Island near Aswan in Egypt in the Nile River. Then jumping past Amon, Manasseh's son, to Josiah, Amon's son, who became king at a very young age, made it an obsession to cleanse Israel from the evil of his grandfather and return it to its honor and standing with God. A great Passover feast took place during this time of cleansing of Josiah. But in a telling Scripture in 2 Kings 23 we learn that it was prophesized that God would bring Israel down because of Manasseh's evil, but it wouldn't occur until after Josiah was gone. Then something startling happens. The Scriptures show Josiah's demand to return the Ark to the temple of Solomon:

"Put the holy ark in the house which Solomon the son of David king of Israel did build; it shall not be a burden upon your shoulders: serve now the LORD your God, and his people Israel," 2 Chronicles 35:3

We see that the Ark was not in Solomon's Temple and that the Levite's must have it in their possession because they were carrying it in proper fashion by the poles propped upon their shoulders:

"they should bear upon their shoulders" Number 7:9

We then see that since the Ark was not in Jerusalem, we had to question, "Then where was it?" Bob went on to tell us that the last part of Josiah's plea in 2 Chronicles 35:3 was for the Levites to begin serving God in Israel once again. That tells us that the Ark of the Covenant was no longer in Israel. This takes us back to Egypt and Elephantine Island. Ancient papyri found at the site show that the priests that were serving there reference as God "dwelling there," or in the "presence of the Lord," which is a direct comparison to Scripture phrasing the same fact when discussing God's presence was with the Ark. But the best biblical evidence that the Ark was in Egypt was during the confrontation of King Josiah and the Egyptian Pharaoh Necho:

"After all this, when Josiah had prepared the temple, Necho king of Egypt came up to fight against Carchemish by Euphrates: and Josiah went out against him." 2nd Chronicles 35:20

Josiah, in his efforts to have God return to the temple, seemed to be blinded by his obsession to return the Ark to the temple. His obsession moved him to attempt to battle against the mighty army of Egypt in an unnecessary war considering that Necho was out to battle the Babylonians.

So Necho, in his passing by and trying to keep peace sends out messengers to Josiah:

"But he sent ambassadors to him, saying, What have I to do with thee, thou king of Judah? I come not against thee this day, but against the house wherewith I have war:" 2 Chronicles 35:21a

But the biggest and most shocking statement delivered by Necho was:

"for God commanded me to make haste:" 2 Chronicles 35:21b

We learn two very startling facts with that explanation from the Egyptian Pharaoh. First, the Egyptian king was telling Josiah that the One and only Jehovah God was giving him the orders. You see, this leader didn't follow tradition and wasn't serving the gods of Egypt's past, but the One God! God had commanded him to march and God was giving the orders. Then, in closing, Necho scolds Josiah with this statement:

"forbear thee from meddling with God, who is with me, that he destroy thee not." 2nd Chronicles 35:21

Necho tells Josiah to back away because God was literally with him on this quest. In other words, Necho knew God Almighty and Josiah would be wise to stay out of their business. But Scripture tells us that the warning from Necho was not enough:

"Nevertheless Josiah would not turn his face from him, but disguised himself, that he might fight with him, and hearkened not unto the words of Necho from the mouth of God" 2 Chronicles 35:22

The Scripture goes on to tell us in verse 23 that Josiah, after masking himself, went out into battle in his chariot and was killed in the confrontation. This then would fulfill the prophecy of God that He would bring down Israel, but not until Josiah was dead. Because of his greed to return God to the temple, Josiah had met these requirements and soon after the Lord brought destruction to the land of Israel.

Jesus: In Ethiopia

The Travels to Lake Tana

So what happened then to the Ark? Where did it go? Well, in approximately the third century, BC, the temple on Elephantine Island was destroyed and the Levites simply disappeared from this place along with Ark. It was at this same time that monks on the forbidden island of Tana Kirkos on Lake Tana claim that the Jewish tribe brought the Ark there. In a makeshift tabernacle, the Ark rested on the high guarded cliffs of this island for nearly 800 years. As we learned earlier, King Ezana became a Christian around 305 AD and proclaimed it to his kingdom. It was also around this time that the king and his soldiers traveled to Tana Kirkos Island and claimed the Ark of the Covenant and transported it and the two silver trumpets of Moses to his kingdom in Axum, where it still rests in waiting today.

I couldn't keep my mind on the rest of the study that Bob was thoroughly laying out for our team. I am thankful that I had my video camera placed up on the back of the benches and filmed the study for future reference. Bits and pieces would get embedded in my mind with the majority passing right by me. I began to flip through the Scriptures to look at Matthew 2 and then to Numbers 10 and then to Isaiah 18. It was like a spinning wind of information between my ears that I had never had before. The knowledge of the Scriptures was coming to light in a way that I had really never processed in times past. It was like God was giving me everything I needed to prepare me, enlighten me, and lift me spiritually for the next day. Listen, I'm one that has to read things over and over again to get it. But all of a sudden it all started to click. Scripture after Scripture started to become clear to me. Even though my head was spinning, everything that I had read, absorbed, and thought about was meshing like a mystical mad house. But what became clearly evident was that it was all making sense. The path of the Ark and its arrival in Ethiopia, and the future calling, were all part of the layout of this incredible journey. But what was still to come? Why all of this speculation? I couldn't quite figure it out, so after the study I stopped Dave Kochis, who had become a great friend during this journey. Dave and I looked at the information; he was having the same spinning feeling to why he was on this adventure, although from an entirely different perspective. I gained a great deal of respect for Dave during this journey. He has a passion and a wonderful, sensible, knowledge that was very relaxing to me. Sherri and I were very thankful that Dave was a part of this journey, and it turned into a friendship that continues today. I felt, which turned out to be the truth, that Dave was re-identifying himself through something personal during our time here. It was later that I found it was an overall cleansing from a personal tragedy in his family that really defined Dave's presence. He didn't even know what to expect, but it all became clear to him as our journey was coming to a close. We were fortunate to gain many wonderful friendships during our journey to Africa. Dave was certainly one of those, as were Steve and Anita, Kathryn, Bev, Tamara, Craig and Meredith, Pam and Michael, the Cornuke's, and others. All had a purpose for going personally and many found that God had another reason for them later. As for Sherri and I, we knew there was still something to come, and I couldn't wait until the next morning to travel on this lake to find out why God had allowed me to travel to this mysterious place.

Chapter 9

THE LOST ISLAND: HOST TO A HEAVENLY SUMMIT

Morning had arrived. For the last year and a half everything seemed to have been laid out without an ending. What was the ending? Why, at my age, had God opened up this incredible life time desire to search for the most sought-after artifact in history to me? Or, was that what it was all about anyhow? I had to believe there was more to come. Just as Sherri had said just four days prior to Bob calling me in 2010 about going on this expedition, "Jim, I really feel that our biggest and best adventures are still to come," I couldn't help but believe that something totally life-changing was going to happen. The dream in the middle of the night about the Ark connected for me with the invitation to accompany Bob Cornuke on the expedition for the truth about the Ark in Ethiopia. The perfect match with the two trumpets on the necklace that the lady in the shop in Gatlinburg, Tennessee handed me as a gift now came to light with the two hammered silver trumpets that were locked away in the hidden vault in Axum. That vault, of course, was next to the church that houses the once-thought lost Ark of the Covenant, that brought me to my knees. And even though during this journey, Sherri and I had developed a great passion for the people in this area of the world, we were also in agreement that our service to reach them through the New Testament gospel for Christ would be a long time future journey. We both also knew that something was still to come. Every aspect of the revelation to me in Tennessee had been put together like a big puzzle with one giant piece in the middle still missing.

We started to look at this a little closer. What was still missing? The dream and two trumpets were complete. The theme of the journey had turned into "When the trumpets sound will you arise and go." This, of course, was in reference to an earlier devotion I gave to the group the night of the revealing of the trumpets in Axum. I shared the purpose of the trumpets to Moses as an instrument to call the people to the tabernacle because God had something to say. But, then I jumped to Acts chapter 8 with the Ethiopian Eunuch and his encounter with Phillip. Prior to this meeting, God told Phillip to *"Arise and go."* Putting those two thoughts together, my devotion to the group was that each of us was called here as if a trumpet sounded in our hearts. What the purpose was may be an individual cleansing or revelation to each individual. But I went on to say that when that trumpet sounds when we return home from this journey, will you take what you were given and arise and go, or follow-up on what needs to be done? But, again, what was this piece that was missing in the puzzle? Or had I gone completely out of my mind? Only time will tell.

As we walked to the shoreline of the mighty Lake Tana, one glance across this massive mystical body of water appeared as though time had stood still with bulrush reed boats carrying fishermen in morning fog off in the distance. Off to the right a variety of birds had congregated on the rocks to begin their morning fishing rituals and the calming waters seemed to passively lay waiting as a lure to those who venture into the waters over the horizon. I was anxious but nervous as I boarded the boat. As we casted off in a small boat, the thought did occur whether this thing was even sea worthy. As the engine started up and we made our turn toward the middle of the lake, I overheard Bob talking with Misgana about whether the waters were low enough for the captain to see the dangerous underwater rock formations. Misgana gave Bob his typical answer, "No problem." I started to review my notes about the missing link to this puzzle. The research team all loaded on and the boat's captain sounded the all clear, and we turned from the dock and headed into a mysterious journey. Bob told us that the trip to the island of Tana Kirkos would take about three and half hours, one way. I couldn't help singing that song from the old television series, Gilligan's Island, about "A three hour tour, a three hour tour." Prayerfully, I hoped for safe passage and not the fate of those stranded in the comedy. The difference, this was not a three hour tour, but a three and a half hour quest to an ancient and shrouded island full of mystery and guarded by monks.

At one point, Kathryn Pierce, the nurse who swabbed Marye for the DNA testing on the mountain in Gondar, Dave Kochis, along with political talk show host Tamara Scott, Steve and Anita Inman, who became our buddies, and Sherri and I took part of the time to put together a strategy to reach the people for Christ on future return expeditions. After all had dispersed to gather in the pounding sun that was barreling down on us, I pulled out my Bible and my notebook and tried to figure out what piece of the puzzle was missing. Then, like thought of embarrassment slapped me in the face, it was plain as day as the key to the missing piece came to me. It was the painting in Pigeon Forge, Tennessee that the lady had to show me, the painting of Jesus and the Ark of the Covenant. If you remember from earlier, this painting was of the Ark plainly in the foreground with the Lion of Judah beside it and then Jesus' face above it. The Lion of Judah was clearly laid out for me as we visited Marye, the Falashian Jew, on the mountains around Gondar. The Lions of Judah were all through the village where Marye lived, made of mud with the Star of David upon their heads. I started to wonder about the Ark in the picture. Was that in reference to the Ark in Axum or was something still to come? It was starting to clear up, but still the reference to Christ was still puzzling me. I tried to convince myself on a couple of wayward thoughts on this matter, but none took anchor. Then I worked in the idea, in convincing fashion, that Jesus in the painting was no more than a reference that Jesus was the keeper of the Commandments of God and became the Law. Unfortunately, my heart wasn't buying that.

The Ark on Tana Kirkos Island

Eventually, after my hyper actively kicked in, I became anxious and put away my Bible and notes and sat my pack down secured on the deck of the boat. I walked around making conversation with members of the team while taking in the beauty of Lake Tana. I even proclaimed that if we had any problems with the hippos, we could throw Anita Inman overboard as a sacrifice. This brought a laugh to the team as we pushed on across the lake. The calming waters were a pleasant foreground to the far distant shorelines and the passing islands. As we passed close enough to the islands, we would see large birds flapping from tree to tree and monks kneeling by the waters fishing or washing their clothes. It was beautiful and actually calming to stare out in to the distance.

After three and a half hours of belting sun, spraying water and the buzzing of our boats engine, Bob tapped me on the shoulder and said, "There it is." I said, "What?" He replied, "Tana Kirkos, right over there." Far off in the distance the silhouette of an island could be seen. As we drew closer, it was visible that it was much different than most of the other islands we passed by. This island had towering cliffs at its base with papyrus surrounding it in the water. As we moved closer, a lone huge sycamore tree sat on a small piece of dry land at its base with something appearing to be jumping around in the tree. Then, way above all of the cliffs, towered another massive tree with two pristine oversized eagles sitting in its top, white heads and black bodies with awe inspiring wings stretched out in the sun. With my background with eagles and Isaiah 40:31 as my favorite Scripture, I felt as though God had given me this as a sign that He was with me and to follow along with what He was about ready to show me. Sherri saw it too, and turned and looked at me as though to say, "Here we go."

As the boat journeyed close to the island, the towering cliffs had to be forty to sixty feet high with what appeared to be a jungle on its top. Bob had told us that the island was guarded by monks who never leave this place. It has been this way for centuries. A group of monks had collected near the base of the island on huge rocks as though they were awaiting our arrival. One monk in particular was dressed in a long yellow robe and round head covering and stood watching as the rest were ready to grab onto the boat. At first attempt the boat hit a series of rocks and the captain yelled out and reversed the engines. The Gilligan's Island thing started to run through my mind again. The captain made a simple move to the right and two of the monks yelled out this time as we plunged forward and hit rock again. One monk stepped down slightly into the water, realizing that these waters were infested with Nile crocodiles and hippos, and waited for the boat to draw near as he guided it to a large rock, made like a makeshift dock.

Even though we tend to think and fear crocodiles, the truth is that the hippos take more lives in Africa than any other animal. Hippos can weigh up to three tons, and when they open their jaws they contain massive teeth with mouths big enough to grab a whole man. Even though hippos are vegetarians, they are very territorial and will attack anything that moves. Hippos even have been known, in these waters, to capsize large boats because they have crossed into their territory. Now, don't get me wrong, one glimpse at a twenty two foot crocodile and you'll realize that these are not waters to play around in. Whether it's a hippo, crocodile, or even a deadly snake, they are all forces of nature and not to be tested.

Now that we were docked, two of the crew leaped from the boat with ropes and threw the ropes off to a couple of monks who tied them to a tree. Finally, three crashes in the rocks later; we were safely anchored to the forbidden island of Tana Kirkos. Bob (or Mr. Bob as they referred to him as) stepped off of the boat to meet the man in yellow who happened to be the Abba, the head monk. After helping our party jump from the boat to the rocks, we all made our way up a broken down opening in the cliffs to the top of the island. The rocks were dry, but very slick due to their use as the only way off the island for centuries. The island is about a mile long and probably only 100 yards across at its widest point. Other than the clearing at the top of the cliff where it seemed to be a mud-daubed building that the monks congregated in, the rest of the island seemed to be dense jungle. The Abba told Bob that the women could go no further with us, but they would go with the monks to a church at the other end of the island about a half mile to the north. As for us, Misgana, Bob, the Abba, myself and the other men on our team began a journey through the jungle. We followed a distinct trail, and at one point I saw what appeared to be a monkey jump across the trees above me. The walk was strenuous as we trekked along the high cliffs above the water below. The temperatures were blaring and the humidity was drenching as we pressed on through the jungle. I think at this point I would have been willing to move onward in any condition. I was on a mission of unspecified instruction and wasn't about to stop now.

On Holy Ground

We arrived at a small clearing in the trees. At first I started to feel the same feeling that I had when I saw the trumpets in Axum. I began to think that the walk had gotten to me and the heat had set in. On our left was a small mud building with a thatched roof, and backed against the jungle, and surrounded on two sides by even higher twenty-five foot cliffs. A bulrush reed boat sat in front of the building on a set of rocks and a monk rested in the shade by the door. I had the sneaking suspicion that someone else was watching us, as well. I looked around without seeing anyone. It was just a creepy feeling that other eyes were on us. Then I saw a monk in the trees behind the building peeking at us. I turned and looked behind and saw the face of another monk, as if they were keeping an eye on us to guard whatever was nearby. To my right was a stone archway with a rock fence and through it was a larger mud-daubed building covered in thatch and bamboo that was the holy temple of the monks. One of them said they called it the church of the Holy Ark. You could clearly hear the eerie sounds of a simple beat of a kebero drum in the building that had no doors, only shrouds hanging down. Along with the drums were chanting that came from the building as many of the monks were inside in prayer. I wondered if this was the same setting when the Ark of the Covenant rested somewhere near this place. I started to walk through the archway and Bob stopped me and said, "We have to remove our shoes." Misgana followed up, "You have to remove your shoes because now you will walk on Holy ground." Within an instant I started to sweat profusely as my mind flashed back to the Scriptures:

"...*put off thy shoes from off thy feet, for the place whereon thou standest is holy ground.*" Exodus 3:5

It was as though I had heard this in a dream or someone else said this to me in preparation for our journey. Either way, it was extremely eerie to hear those words. I put my hand up against the archway and removed my hiking shoes and socks. After we stepped through, Bob pointed at a medieval style cross carved into the wall of the cliff. He had us bring a shovel on the boat, which one of our team carried for this purpose. He told us that the cross was some type of grave marker and possibly for one of the Knights of the

Templar who came to the island in search of the Ark of the Covenant. We waited as Misgana and the Abba stood next to the cross, as we watched from a distance for fifteen minutes or so. I'm a very anxious person, so I headed for the temple where the sounds of the drums continued to pound out. I pulled the shroud to the side and glared into a very large dark room. The floor was covered in palm and banana leaves and Bob slipped up behind me and warned me not to step inside because, even though it's a place of worship, it is also a haven to millions of fleas. He went on to tell me that these fleas aren't the little pesky fleas we have in the states, but a larger and more potent flea whose bite won't go away for awhile. I wasn't in the mood for itching for the next week, or to become a tasty meal for a batch of waiting fleas searching for fresh meat instead of the hardened calves of the monks, so I backed off and continued to watch the negotiations of Misgana and the Abba. Finally he returned over to us with a plain and simple "No." Bob said, "All we want to do is dig up the grave to confirm that this is who we are told it is." Misgana once again said, "No, the Abba says no." The explanation was to say that since this was considered Holy ground, we were in no way going to defile the area by digging up the remains of a body to prove a point. So we sat the shovel along the fence and turned to continue our trek up the cliffs.

The Tabernacle: The Place of the Ark

As we turned to make our way around the temple, the tingling sensation in my body started to intensify. Maybe the drums slowly beating inside the building, the chanting of prayers and the sun were starting to get to me. Or, at least, that's what I thought. On up the cliff another hundred yards, where the rocks started to become very narrow, only 25 feet or so before dropping off forty feet below into the Tana waters, the feeling, just like in Axum had taken me to my limit, and I dropped down to my knees on the rocks and I looked down. The flat slab of rock was somewhat covered with dead grasses and small stones and I began to quickly brush them away. It was almost immediate that I saw a small round indention in the slab measuring approximately two and a half inches in diameter. As I kept looking, I saw another about a foot and a half across extending from that one measuring the same two and a half inches. Bob, in the confusion, said, "Jim, what are you doing?" "I don't know." I said. "But something led me to this spot like I had been here before. I don't know what it is but something is or was here." Bob looked down, as Dave and Steve stood by for the answer, and said, "How did you know that was there?" I told him I didn't have a clue what was there but something was, and I pointed to the indentions in the stone. Bob said, "That is the exact spot that the Holy Ark of the Covenant rested for nearly eight hundred years. Those are the indentions of the feet of the Ark to keep it stable." My thoughts were running wild. How and why did God lead me this place? I handed my camera to Steve Inman and asked him to shoot video while I attempted to find the other two markings or foot-holds for the Ark. As I knelt in the place of the Ark I couldn't help the tears that were freely flowing down my face. There was something about that spot, that place, and I wasn't sure what it was. If this was truly where the Ark rested, there had to be two more and their measurements would have to be close to the measurements of the Ark depending upon their placement on the bottom. I continued to pull weeds and throw them out to the side and there was another mark approximately three and half foot from the upper indention. But, where was the fourth? Finally, I pulled up a thick piece of grass that was covering a portion of the stone and there it was. About a foot and a half across was the fourth marking for the foot of the Ark of the Covenant. I now felt the statement that I had heard just a little while earlier was flashing like bright lights through my mind, "You now walk on holy ground."

Bob told us to step over and look at the large holes carved into the rock. He pointed out, "We have one here and another over there, (pointing) about thirteen and a half feet from this one. These were where the poles were set for the tabernacle's holy of holies to cover the Ark. A small tin building sat on the cliff-side just a few feet away and inside were two more pole settings that had broken off the cliff; they put them in the shed and also the stone carved bowl that the high priest would put the blood of the sacrifice in for the atonement of sin in the presence of God. The Abba demonstrated the sprinkling of the blood with his horsehair whip. He snapped first toward the top where the Ark would have rested and then to the ground below. I asked, "Why on the ground?" Misgana replied "That is where the soles of God's feet would rest." Then I remembered the passage where God says:

"And he said unto me, Son of man, the place of my throne, and the place of the soles of my feet, where I will dwell in the midst of the children of Israel for ever, and my holy name," Ezekiel 43:7

The Ark of the Covenant was the throne of God on this earth and in that respect the blood was sprinkled on the throne, or to be more specific the Mercy Seat, and then at the base of the Ark on the ground where God's feet would literally rest. That is the exact place where I was on my knees digging for the post marks for the Ark.

Where Jesus Knelt with His Father

After inspecting closely the sacrificial stone bowl and pole indentions, we made our way on up the cliff. I glanced down at what appeared to be another indention in the rocks on the edge of the west side of the cliff. When I went back down to my knees, I grabbed hold of the edge, where the indention was, and it broke off into my hand as I dug my fingernails into the rock with my left hand to keep from falling off. Even though there was a ledge just below, I still didn't want to run any risk of tumbling on down to the waters below. If the fall didn't kill me, the inhabitants of the waters would definitely take care of me. I threw the piece of rock into my bag, which had fallen off my shoulder and was hanging over the cliff, and pulled myself up with a clump of grass. After I lifted myself back to my feet, I decided to move onward to a pair of large boulders on the left side of the pathway. Between the boulders, smaller watermelon-sized rocks were stacked high between them. One of the monks said that is where the Levite priest who brought the Ark to Tana Kirkos was buried and his remains lay under those rocks still today. Think of it, the body of the actual Levite priests who came from Aswan, in Egypt, with the Ark of the Covenant, lay in a very humble grave just inches in front of me. After the others in the team passed by, I climbed up the backside of one of the boulders and looked down inside the tomb. Although I couldn't see anything inside, I'll trust their claim that this was also a very holy site in the trail of the Ark. I then went up another thirty foot and the monk once again said that this was the ledge where the Ark was lifted from the water, some forty or more feet below, to this cliff where it rested for nearly eight hundred years.

I stepped back for a second to take in all that was around me: Holy ground, the Ark's resting place covered by a tabernacle, the high priests of the Levites buried between two boulders, and the lifting site where the Ark was hoisted for safety on this incredible remote island. It was almost overwhelming to take it all in, but even more incredible was to realize the tremendous divine history with the shroud of mystery that surrounded this particular land I stood upon and the presence of God in this spot. But for some reason, the Ark and the thought that we were passing upon holy ground just would not leave my consciousness. I didn't know what was going on inside head. It just seemed like there was more. There had to be more. There was something that I was still missing. There had to be something that I had missed or just had not found in my search as of yet.

So I made my way down the cliff again to the site where the Ark had rested. Many of our team had already started to make their way back, but I knew something else was left and possibly covered up. Where was it? What was it? I didn't know. Again I went down on my knees where the Ark had rested and the Abba said something to Misgana, and he turned to Bob and said, "They tell me that where you are kneeling is where Jesus knelt." I turned around my head, with my knees still planted firmly on the rock and said, "What did you say?" Bob, with his arms crossed and head tilted, answered, "The Abba said that when Jesus was a boy He knelt where you are with His Father at the Ark. The Abba says to come with him." Tears began to roll down my face while still being in a bit of a shock. I turned to Steve holding my video camera and described the four post marks again and the revelation about kneeling in the spot of Jesus while really unclear to what I had just been told. Confusion rattled inside my head as I sat there for a moment on the exact spot where God Himself rested His feet and now something to do with Jesus actually coming to this place to pray on this slab of rock as well. If what I heard was truth, the place where I am resting on my knees would actually be the only place on earth that both God the Father and God the Son would have ever met. Think about it, a remote island guarded from the world and inhabited by only monks who are chosen for their purity and glaring with their dedication to faith. Could this be possibly the most Holy spot on earth or at least a rival to the top spot? I collected my thoughts and made some notes in my journal about the resting place of the Ark. Then Dave grabbed my hand and pulled me up as we followed the Abba back down the cliff, past the temple, and through the rock archway to meet up with those who had already ventured back and missed the shocking announcement. God was in the midst of opening up a revelation and I prayed that I was prepared to take it all in without losing any of it.

MARY'S RESTING PLACE

After clearing the archway and leaving the place considered to be "Holy ground," I put my shoes back on and we followed the Abba to the small mud-daubed building nestled between the archway, two higher cliffs, and the jungle. As the Abba went to collect the sets of keys to unlock a very heavy wooden door with three large extremely old locks keeping the contents concealed from the world, the thought of Jesus kneeling here may be the big missing piece of the puzzle. While the Abba was gone, Misgana said to follow him up the higher cliffs. As the temperatures were rising and the sun was at its peak in the sky barreling down on us, Misgana led me up the cliff to the highest point, to the top of the rock. He pointed to a spot on the rock that had a small marking on it and said, "The Abba will show you more, but this is the place where Jesus' mother Mary would go pray. She could see Jesus with His Father at the Ark (pointing to the other cliff where the Ark rested) over there. Down there (pointing at where Bob waited for the Abba at the building housing the mysterious answers now rested) is where Jesus would play and they would eat." I put my head down to regain my composure, and then looked back up to Misgana and asked, "So, what you're telling me is that Jesus Christ was actually on this island when he was a boy. He was really here?" "Yes," he said. "The Abba will show you down there."

Could this actually be the place where some two thousand years ago Mary sat, rested and prayed while watching her Son Jesus with the Ark and God the Father's presence upon it? If this is the place, it appears to be the perfect setting for such a truth. From this point, I could see everything: an undefiled piece of rock and earth that was created by God to be the site from the beginning of time to host a heavenly conference. I can see the other side of the island where the women had to stay back. I can see for miles over the lake to the non-inhabited shoreline of the mainland. Islands are in all directions and could be seen in a variety of shapes and sizes. But the most important sites in the vision from this high rock cliff is the place where two of the Godly trinity met here on earth. Never before has there ever been such a place below heaven that these two have met. Never will there be such a place again according to the Scriptures. I can only imagine what Mary thought as she gazed over the trees to the tabernacle. This young woman and young mother had not only given birth to the Son of God but she was witnessing a supernatural meeting between her Son and His Father.

I tried to take it in and position myself where I felt that Mary would have rested and prayed. Then it clicked, I felt like I was beginning to get the purpose behind this power summit. But realizing that more needed to researched, I grabbed my Bible from my backpack and quickly opened up to Luke Chapter two. I remember from reading this chapter many times over the years, one line of Scriptures that, for some reason, just resurfaced to my brain. I have read this before but it never materialized into anything I felt I needed to adhere to because of the main stories that bookend this line. Of course, preceding this single line is the Christmas story and then it is followed by Jesus' disappearance and Mary and Joseph going back to find Him in Jerusalem teaching in the temple. But then you always hear about testimony of people saying they have read Scripture over and over again, but not understanding it or not paying attention to it until God needs you to. I think that is what may have happened here. I have read and passed it by many times before. Luke says in his gospel:

"And the child grew, and waxed strong in spirit, filled with wisdom: and the grace of God was upon him." Luke 2:40

This was after the Holy family had returned from Egypt and all of the customary rituals had been completed. Could this be the clue to that leads us to this remote island in

Ethiopia in the middle of Lake Tana? We may be resting in the very spot where Jesus and God the Father met in order to fulfill this Scripture. This Scripture reveals to us that Jesus, the man God, was brought to be strong in His Spirit and filled with the wisdom and knowledge to prepare Him for what was to come with God's grace upon Him. It started to sink into my simple mind that God might be clearing a path and slowly opening me up to a revelation about Luke 2:40 and offering the "when" that Jesus received His wisdom from God, and "where" that this grace was placed upon Him. Tana Kirkos Island was possibly the place of a great meeting. This could very well be the place where The Son was filled with all that was needed for what was to come.

As I stood there with watery eyes and a spiritual feeling washing over me, I turned around to look at the awe-inspiring view over Lake Tana and started thinking to myself, "Could this possibly be the ending stop to the Holy Family for Jesus to meet with His Father in privacy to receive instruction for an entire lifetime of direction and all knowledge and strength to face the many events to come in His life?" I stared out over the lake at a nearby island which contained an ancient ruin of a crumbling rock building that was surrounded by jungle trees. Then, off in the distance was a very deserted but lush green coast line of Lake Tana. Misgana, who had wandered farther on the cliff, returned to me and said, pointing to the mainland, that is where Joseph (Jesus' earthly father) stayed and farmed the ground for food and money. I asked, "How can this be? The Bible says nothing of Jesus coming here." He replied, "There's much in the Bible that it doesn't tell you. But, you will see (pointing back down to the building) that it is true, Jesus was here. Jesus was here as a boy with His Father."

Jesus: In Ethiopia

The Island Treasury

Tana Kirkos Island in one of Ethiopia's largest lakes, Lake Tana

Author Jim Rankin kneels at the exact place that the Ark of the Covenant rested on the island and where Jesus knelt with His Father

The harness for the breastplate of the high priest from Solomon's Temple

The gomer, or possibly the lever, from Solomon's Temple housed in the treasury on the island

I made my way down the cliff, while climbing over drying coffee beans, and went straight to Bob. "Bob," I said, "I need to talk to you for a minute." He answered, "Just hold on a minute (as the Abba was unlocking the three locks on the door). I told him that you are a holy man from America and he has agreed to show you the book." I said in surprise, "What book?" He came back to me in a near whisper, "The book that holds the history of Jesus' visit here that he told you about up on the cliff." Thrill, adventure, anticipation and confusion were all encompassing me at that very moment. Not sure whether to cry, argue, fall on my face in prayer; I just stood there with an overwhelming Spirit over me, dazed as the Abba opened the door. I followed Bob in as the rest of the team came in behind us.

There was only room for a few as we stepped beyond the doorway of this musty mud building. To the right was shelf after shelf of ancient scrolls, books, and papyri. To the left lay upon makeshift wood and bamboo shelves, what appeared to be a treasure trove of dusty artifacts that only the Abba knew for sure what they were. The Abba handed Bob a cage looking piece of pounded metal. Bob handed it to me and said this is the remains of the breastplate harness of the high priest for the tabernacle atonement ceremony.

The Abba pointed and mumbled some words as in reference to the piece. Bob, as if he understood what he was trying to say to us, went on to explain that the cloth with the twelve stones of the twelve tribes of Israel would have be laid over and attached to this breastplate.

Then the Abba handed Bob a large aged bronze bowl. This was the bowl where the blood would have been placed for the sacrificial service in God's presence during the Levite priests' atonement ceremony in the Holy of Holies. They called it a gomer, but it also had a striking look to the description of the lever that the high priest would wash himself from. A crumbling bronze stand, next to us and resting on the ground, was the piece that the bowl sat upon. Due to the age and weight of the bowl, the stand had nearly collapsed. Then the items just kept coming, one after another. Next were meat forks with almond bud designs on the top. These were the forks that held the sacrificial meat for the burnt offerings and the almond was the sign from Aaron, Moses' brother, serving as the high priest. It was Aaron's almond staff that God made to bud and flower that was included inside the Ark of the Covenant. This, again, was a sign of these items truly coming from the time of Moses and continuing on to Solomon's temple and then to Tana Kirkos.

Jesus: In Ethiopia

Aaron's symbol of the almond was just another piece of the puzzle that didn't have to be shown but was in a confirmation of what we were seeing as truth to the claims of the monks. There were ancient shofar horns, incense burners, additional meat hooks for hanging the meat and a variety of other items. I've read accounts of the Egyptians opening up crypts and finding a slew of ancient and historical artifacts. This was much the same, but we didn't have to become tomb raiders to find these artifacts. It was like a biblical museum in a hut made of mud covered with a thatched roof.

There had to be some merit to all this. There was little to no metal at all on this island. Everything for the most part is made of wood, reed, or rock. Where did they get these things? Are they truly from Solomon's temple? Misgana, translating for the Abba, said, "These are all from Solomon's Temple and came to the island with the high priest who brought the Ark here." This, of course, is the same high priest whose body lies under the rocks farther up the cliff-side. I had to ask, "Then, why are they still here? The Ark and the trumpets of silver are in Axum, so why are these pieces still here?" The Abba, through Misgana, explained, "King Ezanus came here with his army and said that the Ark belonged to him and he wanted it in his kingdom in Axum.

They were now Christians and believed in Jesus so they did not take anything for the sacrifice because they didn't need it there." I questioned, "So, all the blood sacrificial items like the breastplate, meat forks and gomer were left here." Bob said, "Yes. They didn't need them anymore because they were Christian."

It all started to make sense now. The Ethiopian King only took with him the needed pieces of the treasury. The trumpets of Moses traveled with the Ark everywhere it went to announce its arrival. Then, of course, the Ark itself went to Axum and they claim it continues to rest there today in the St. Mary of Zion church. Our team appeared to be like detectives looking and investigating everything. A couple of times someone would reach for something and the Abba would raise his hand and shake his head no. In other words, "Don't touch!"

THE BOOK AND THE REVELATION

The 1900 year old book sharing the history of Tana Kirkos Island with Jesus and His Father together next to the tabernacle along with Jesus and the Holy Family crossing Lake Tana to meet with His Father

Page Prochorus sharing the words of John as in the Bible in Revelation 22

While all this was going on, I turned to Bob and said, "What about the book?" Bob leaned over to the Abba and Misgana and whispered something. Misgana motioned for the men to leave the building as the Abba turned to rustle through the stacks of scrolls and animal hide books. Bob, Misgana and I stood by as the Abba seemed to have found what we asked for. He turned toward us, and I took a deep breath and a big gulp. The pages were stiff to this book as it was written in some type of ink nearly 2000 years ago. I wasn't sure whether it was berry juice or some other ink-like liquid on a thick animal hide. Then he stopped on a page near the end of the book. The Abba pointed to the writings on the left side of the book and said, through Misgana's translation, "This Bible tells us that Jesus traveled here from Egypt when He was a boy to be alone with His Father." I said quickly, "Jesus came to this island?" The reply was, "Yes." I turned to Bob and asked, "How can this be? There are no writings of this at all." Bob said in very specific words, "It appears as though there are now." This sent shock waves down my spine, and I trembled. My face was flush and I instantly felt both drained and exhilarated. I turned back and belted out, as if to quiz the Abba, "How long was he here?" He went on to tell us that Jesus was here three months and ten days, and He prayed by the Ark sometimes night and day. Then his hand slid over to the right side page that contained a beautifully drawn picture in blue, red and black ink. It was a picture of a small boy, "Jesus," the Abba claimed, "His mother and two others on a small bulrush reed boat being rowed out to the island." Then the following pages were opened to other incredible writings.

I asked if we could take the book outside so we could get a clearer photo of the book and its writings. The Abba agreed and we moved to the outside of the building and under the bamboo and thatch covering as the sunlight exposed the beige-like pages of this Bible. We could see it much clearer now. The others on the team appeared just as thrilled as I was. How could we not be? Bob and I were the first of the outside world to be shown this book and the first of anyone, other than the monks who have guarded these secrets since ancient times, to be shown this incredible writing of significant biblical history in nearly 2,000 years. Little did I know the path to completely reveal what was written in this book was being cleared. Quickly, I was going to be presented with the clear vision that this was one of those mysterious that God was opening up for the world to know.

Chapter 10

ALONG THE PATH TO TANA KIRKOS

After studying the book, the scripture inside, and the trail of the Holy family, it appears as though the journey into Egypt was only the start to a much bigger adventure for Jesus. As we follow Him through Egypt, all of the Egyptian writings have Him stopping in the town of Muharraq. To me, that never made sense. Why would they retreat into the southern regions of Egypt when Herod's army didn't even know who they were looking for? Why would they travel so far to the south when Herod's soldiers wouldn't have dared to travel into the Egyptian desert? The Scripture tells us that;

"Herod, when he saw that he was mocked of the wise men, was exceeding wroth, and sent forth, and slew all the chil-

Jesus: In Ethiopia

dren that were in Bethlehem, and in all the coasts thereof, from two years old and under," Matthew 2:16

When Joseph packed up the family and headed for Egypt, it was an escape from Herod's wrath in Bethlehem, not in Egypt. It was obvious that Herod had no idea who Jesus was. So to compromise for his lack of information from the wise men, he ordered the slaughter of hundreds, if not thousands, of innocent young children and dumped the bodies in what is today known as Herod's Catacombs. Besides, his men would not be a welcomed group in this African land. If they would have attempted to follow the Holy family into Egypt, they wouldn't have followed very long into this blistering hot desert region.

As we saw before, Jesus, along with Mary and Joseph, traveled almost in a looping pattern in the northern regions of Egypt before connecting on the Nile River and heading due south. My investigative nature kicked in as I stood there on a remote island, guarded by monks, and accentuated by a clear biblical spiritual feeling surrounding it. I immediately went over and sat down on a rock under the reed boat resting on the cliff, and swung my back-pack from my shoulder to the ground and unzipped the large compartment in the front.

Once again I grabbed my Bible and looked up Matthew chapter two and then a couple of other Scriptures telling of the Messiah in Egypt. I checked out the reference to Isaiah 19:1 as the idols fell upon the presence of the coming Messiah through Egypt:

"Behold, the LORD rideth upon a swift cloud, and shall come into Egypt: and the idols of Egypt shall be moved at his presence"

This tells us that when the Lord moves through, like a swift cloud through Egypt, that the idols of that nation shall be moved or fall in His presence. When Jesus moved through Egypt, all along the path, the idols and temples of pagan gods fell to the ground. Every location that describes the trail of the Holy family includes these facts today in these locations. This all referred me back to Isaiah again:

"Woe to the land shadowing with wings, which is beyond the rivers of Ethiopia: That sendeth ambassadors by the sea, even in vessels of bulrushes upon the waters, saying, Go, ye swift messengers, to a nation scattered and peeled, to a people terrible from their beginning hitherto; a nation

meted out and trodden down, whose land the rivers have spoiled! All ye inhabitants of the world, and dwellers on the earth, see ye, when he lifteth up an ensign on the mountains; and when he bloweth a trumpet, hear ye. For so the LORD said unto me, I will take my rest, and I will consider in my dwelling place like a clear heat upon herbs, and like a cloud of dew in the heat of harvest. For afore the harvest, when the bud is perfect, and the sour grape is ripening in the flower, he shall both cut off the sprigs with pruning hooks, and take away and cut down the branches. They shall be left together unto the fowls of the mountains, and to the beasts of the earth: and the fowls shall summer upon them, and all the beasts of the earth shall winter upon them. In that time shall the present be brought unto the LORD of hosts of a people scattered and peeled, and from a people terrible from their beginning hitherto; a nation meted out and trodden under foot, whose land the rivers have spoiled, to the place of the name of the LORD of hosts, the mount Zion." Isaiah 18:1-7

This Scripture describes Ethiopia perfectly. Remember, at the time of Isaiah, Ethiopia was a building empire. The Scripture clearly points to an Ethiopia that has suffered and has changed little over time. Both the lands and people are suffering. It talks about the mountains and how a sign will be seen coming from them. In Ethiopia, these northern regions are the mountainous regions of the country: Axum, Gondar, Lalibela, Barhir Dar and yes, the island of Tana Kirkos. It describes them traveling in bulrush or reed boats and how one day a very special *"present"* will be brought from Ethiopia to the Lord of hosts, Jesus, when He takes up His new kingdom on the throne in Jerusalem. We know that gift as His throne, the Mercy Seat, in which the priests in Axum claim they have, and the entire country believes it as well.

    I then glanced back at Luke 2:40 one more time. I quickly made notes in my journal and then trotted back over to get one more glimpse of this book before it was put away. The Abba closed the book and alone went back into the building; the book seemed to have disappeared into the shelf with the other old manuscripts and religious writings. I felt yearned to spend the day just holding and meditating about this book. Is it possible that this is untold history? Possibly another secret hidden away in Ethiopia for centuries until it was time to be brought forth? Could it be a historical account of Jesus in a preset meeting with His Father to prepare Him for His ministry upon this earth? If discernment can be trusted, this was exactly the feeling I was getting. There was a supernatural sense about this place. It was a feeling that I had never experienced before, and it felt as though my thoughts were not coming from me, but rather given to me to document and prepare for the world to receive.

TIME TO LEAVE THE ISLAND

Even though I felt I could have stayed there for days on end, we made our way back through the jungle to the cliff opening where the ladies on our team were anxiously waiting for our return. Before we left the island, there was still much to do for the monks who had assembled to join us. Many of these men were very old with many ailments due to lack of medical attention. Since they don't leave the island, they don't get the proper care they need when illness or pains arise. After giving them ointments and pain medications, Kathryn and Dr. Sue took cheek swabs of the two older monks. These men claim to be from the Levite sect, and we wanted to get some of their DNA in order to try to squelch any doubt that we or anyone else may have.

We also spent some time mingling with the monks, exchanging gifts, and taking part in the custom of drinking coffee. I have to admit that I am not a coffee drinker and this concoction confirmed the reason why. It was simply awful, but who am I to judge. Sherri took off her bracelet that I had given her for Christmas that shares the plan of salvation through Christ and gave it to one of the women on the island. These women come over from the mainland to the top of the cliff to cook for these monks next a mud and thatch building. Sherri tried to explain to them that it was a bracelet telling the life of Jesus Christ and that He died for our sins. They shook their heads as if they understood.

Some of the medical supplies we carried with us were used on some of the monk's eyes and a variety of skin irritations that they had. We gave horsehair fly swatters to a couple of the men, as well. After some final traditional Ethiopian handshakes and goodbyes, we began to make our way down the slippery cliff side once more. Before Sherri and I made our way down, we walked to the other edge of the cliff. It was a fifty foot drop straight down and we could see birds nesting in their little mud homes and the waters of Lake Tana splashing up against the rock walls. I glanced around the island one more time and glared out over the lake from that side. I said to her, "Can you imagine or even contemplate what took place on this island? God came down from Heaven right here and met with His Son Jesus on this place." She looked at me puzzled and I told her that I would share the whole story with her on the boat.

As we gazed out over the lake, tears welled up as I was praying to myself for God to show me something to confirm that what I had seen here was the truth and what I was supposed to do with this revelation. We turned and I looked down on the ground and noticed, buried in the grass, a cross made out of papyrus nestled in the dirt. This was the same type of plant used to make the boat that Jesus traveled in to this island according to the ancient book that the Abba had shown me. I reached down, picked it up, and slipped it into my Bible secured in my backpack. Later I would find that the place where I had quickly inserted the cross was Isaiah 18.

To me, it was a final confirmation that I had come to search for the Ark of the Covenant, but somehow had ended up in the middle of the beginning of the road to the Cross for Jesus. The knowledge of His mission may have started right here on this island. This small cross made of bulrush may be God's way of saying that, "I sent you here to reveal another story to the world that it was time for them to know." A peace came over me as I put my arm around my wife with tears in my eyes. At this moment I knew it was time for me to go, while also realizing that this wasn't the last time I would set foot on this island. I would be back here some day because I felt that God had more to show me when I return again.

We made our way to the cliff side with the rest of our team and headed back down to the water. I turned and shook the hand of the Abba, once more. After we boarded the boat and shoved off, I took one last glance at the island with the two large white and black eagles standing guard and perched in the tree above the cliff where Mary rested and prayed. We waived to the monks on the rocks who made their way down the cliff to send us off as the boat's captain made a swift left turn, and we were on our way back. I was like a frazzled little kid as I attempted to tell Sherri about what we had just seen and heard "on the Holy ground." She was in awe, as were a couple of the other women, as I went through what we just had witnessed on this island piece by piece. I pointed up to the eagles and said, "That's the spot where Mary prayed and watched over Jesus. Right over there is where Jesus knelt in prayer with His Father at the Ark of the Covenant, where the tabernacle stood. The high priest that brought the Ark to the island is buried between those two rocks and that cliff is where the Ark was pulled up from these waters." Sherri and Anita responded, as did Kathryn, "Jesus was here?" I said to them, with Dave and Steve standing by, "Yes, Jesus came here to be alone with His Father." We can only imagine what that period must have been like, what must have been talked about and what came from this meeting. I believe these revelations are perfectly aligned with Scripture and fulfill many prophetic declarations. To think, it all started right here and the world has been hidden to it for two thousand years.

I jumped off the bow onto the walkway along the side cabin and headed to the covered awning over the back of the boat and began to research. As the motor pushed us farther on across the lake, I couldn't help but to think about the path that Jesus took to get onto this island. Did He come from the north or from the southern part of the lake? I wonder how long it took them to get there. It was like questions upon questions were reeling through my head, and it was time to write them down and answer them. I had studied Jesus' trek into Egypt just months earlier. But now it was more than just a study, it had become a mission. It was an Indiana Jones-type adventure, a Sherlock Holmes mystery, and a James Bond thriller all wrapped up into one. This surely did become a mission. Not just any old mission, not a fictionalized adventure, but one of revelation to the world.

THE HOLY APPOINTMENT

As I continued to look at Jesus' trip down the Nile to Deir Al-Muharraq, I started to really understand that this was no coincidence. Herod's soldiers were not following them that far into Egypt by any stretch of the imagination. The Holy family was using the Nile as a GPS to follow the mission that God was sending them on. They were on a journey to a Holy appointment between Jesus, the Son of God, and His Father. What was Jesus, the King of Kings, doing on a remote island in the middle of a giant lake in the northern highlands of Ethiopia? Why was He alone, except for His mother on top of the rocky cliffs and the monks that guarded this island somewhere off in the distance? Or were they? Could this be a coincidence that the Ark of the Covenant was resting here at the same time? Questions were streaming through my head faster than answers were coming. Then, it all started to come together and make sense, like a Rubik's Cube when the little squares start to match-up.

Best-selling author, Paul Perry, had done a tremendous amount of research, written a book *Jesus in Egypt*, and hosted a documentary *Jesus: The Lost Years* about the Holy family's trek into Egypt. After an interview with Paul for our "Adventures in Truth" article in Cross Point Magazine, I shut off the recording devices and asked him, "Why do you think Jesus and his family traveled all the way down to Deir al-Muharraq before getting the instruction to turn around and head back home?" He had no explanation or theory on that question. You see, Paul's extensive study was only on the topic of Jesus in Egypt, and he is the best of the best on this subject as he traveled the entire trail of the Holy family for his research. During Paul's time with the monks and priests in Deir al-Murharraq, he was told that after touching the rock that Jesus slept on he was supposed to pray for something that he really wanted. Paul simply prayed for some confirmation that this trail of the Holy family was true. He received that confirmation through a flash of light over the temple during a photo opportunity that he took later that night (*ref. 5*). Knowing that I could trust Paul on his answer, I continued, "How do you know this was the last place they traveled to? Even though Joseph was told to come on home here, how do we know this was the last place?" Paul replied, "It appeared the trail ended here and we just assumed this was the end." Boom! The word "assume" was what turned my investigation into high gear. For Paul, his mission was complete because his study was centered on Jesus during His time in Egypt. We were taking this one step further, because our adventure was turning to Jesus in Ethiopia, which until now, had not been revealed to the outside world. Was it a step of faith? Absolutely! But it was a step where every possible rock and road was placed on the path for us to follow.

This was more than just a vacation for the Holy family; it was a hard, merciless trip down the crocodile and hippo infested Nile River through Egypt. The Nile continues south out of Egypt into the Sudan. During my study, I realized that the Nile River even went a little farther than I had first thought. I found that Lake Tana (Ethiopia) through the Blue Nile Falls dumps right into the Blue Nile River, traveling into Sudan, meeting up with the White Nile and forming into the Nile River. Lake Tana is the source to the Nile which helps to explain the route for the Holy family. They traveled to Deir al-Muharraq in Egypt, along the Nile River, straight down to the Blue Nile and right to Lake Tana where they loaded up in a reed boat and traveled to the island of Tana Kirkos. Later we find that the family even took in some other locations in Ethiopia, which later became very significant in the foundation of the country and even the housing of the Ark of the Covenant.

There were still many questions to answer, though. I had a strong sense that most of what I was thinking would be confirmed in the pages of the ancient book that the Abba showed me on the island. If this book contained in it what I felt, then we were on the verge of a hidden secret. Not one that will change the world in Christianity, but one that will offer up a whole new look into a once thought, unrecorded time period in the life of Jesus Christ. Many questions still were unanswered, but I felt that if God's hand were in it, then He would also open up the doors and clear the paths to make the answers clearer.

Chapter 11

ANCIENT WRITINGS: JESUS AND THE FATHER MEET

As our boat drew closer to the shoreline, we could see young boys pushing themselves along with a long wooden pole in the water while standing upon a huge reed matt. It almost looked like a floating carpet on top of the water. Another boy passed by, without even a stitch of clothing, in a papyrus boat fishing, while another group of four men pushed their large papyrus matt along in the water with the help of a long wooded pole. Huge white and black cranes along with a variety of other birds rested in the trees while the captain yelled out something to us in Amharic. One of his mates came running out of the cabin and pointed to two large hippos in the water. We slowed as we came closer to these two awesome giants as they watched us closely as we entered their territory. Again, hippos are very territorial, but these two didn't show much interest in us at all. I was amazed by the young boys in the boats as they passed by since Hippos are responsible for more deaths in Africa than any crocodile. But to these boys it was like business as usual. The same held true for the hippos as they glared over at us like we were the ones on display.

We turned in closer to our dock to see dozens of citizens of Barhir Dar standing on the rocks next to the lake washing clothes and bathing. Cormorants (a fishing bird) stood on rocks protruding from the calm waters of Lake Tana with their wings spread and soaking in the sun's rays. Pelicans and a variety of other birds swooped from the sky to take in some afternoon fishing as the sun pelted down on the waters of the mighty Lake Tana. Men were waiting for us at the dock as the captain moved our boat into position to throw the ropes in for tying. We again jumped from the boat onto the floating dock and made our way back to our rooms. The information we had just gathered was still reeling in my head. We went back to our room, and I sat down on our bed and tried to go over everything again with Sherri. She couldn't believe what she was hearing. The revelation to her was as powerful as the impact on me. It was information overload, and I needed to take a break from the thought for awhile.

We decided to go out and walk around the town square of Barhir Dar and take in some of the sites. We came upon a small group of merchants with their wares on blankets off in the grass, just off the roadside, and we started to realize we were being pushed by a quickly building crowd. Little did we know we ended up walking right into the middle of a Muslim gathering near a mosque. Within seconds it seemed as though we had been filed in and then pushed toward the mosque by thousands of Muslim followers. As we attempted to make our way away from the crowd, the trend to swallow us into the mob became somewhat overwhelming. I grabbed Sherri's collar and pushed her forcefully toward the busy roadway. I told her to not stop for the traffic; we needed to get into the road to get away from the push of the crowd. Suddenly we were in the road with the small taxis buzzing and beeping by. We jumped to the square in the middle of the circle and caught our breath. With some fear of being recognized as an obvious Christian in the mob and being in the wrong place at wrong time, we crossed to the far side of the street and picked up a few last trinkets. This commotion was too much after our experiences earlier in the day, so we decided it was time to head back and wash up for our final dinner together as an exploration team.

The mood was festive, but the talk at the tables centered on what happened earlier in the day. After dinner everyone wandered and relaxed around the property. Some pulled up a chair and sat in the moonlight gazing out over the calm waters of the lake for one last glance at this incredible place. Others went to their rooms to prepare for the journey the next day and to get a good night's sleep, Sherri and I walked around the facility. Standing on the shore staring out and thinking what it must have been like when Jesus arrived here, we couldn't help but try to visualize a papyrus boat being led out across this huge expanse of water to take a young boy and His mother to a lost island. It was an island that was pure in its beginnings and remained that way to ensure the opportunity for The Summit between God and His Son.

Many people believe that this portion of Ethiopia was actually the site of the Garden of Eden. It would honestly make sense that God would bring His Son to the place where He created mankind in order to share with Him how He was to save mankind from their sin. Think about it. The Garden of Eden was the place where God came down to walk with Adam. It was the place where the first sin was committed and the place where God would share the ministry of how His Son would wash away the sin that started there. There was so much to ponder upon. After meeting with Bob one more time, we made our back to the room, packed our bags and prepared for an early morning departure. Bob could see that the weight of the tremendous information was taking its toll on me. But just as it had done for him, if it was going to be something that God wanted revealed, then it would be something that God would lay the foundation for us to walk upon.

Morning arrived and we hurriedly threw our duffle bags on top of the vans and made our way to the small airport. Even the small Ethiopian airports house people trying to sell their handmade souvenirs. Then there are others who have hidden away older and sometimes ancient artifacts. You don't question where they came from, if it is of interest to you, just negotiate, put it in your bag and walk away. After waiting an hour, our plane arrived to take us back to Addis Ababa. We walked out on the tarmac and boarded the turbo prop plane for the last time on this journey. The plane barreled down the runway and with a quick pull back on the throttle, the captain lifted us up into the sky, I looked down on Lake Tana as we passed by its shores. My thought was simply, "Wow!" A place of history that has been hidden from the world for two thousand years had just revealed itself to me. Sherri could see the joy on my face and, in some way, I think it made her proud to know that we were entrusted to share this revelation with the world.

Once we landed in the capital city, it would be nearly 40 more hours before Sherri and I would finally lay our heads on our own pillows in Ohio. I didn't want to leave Ethiopia but realized I had much more excitement to come. It was an evening flight out of Addis Ababa and a couple of our team members, Steve and Anita Inman, were detained when they attempted to get their video equipment through the government check point. They didn't get to leave with our group but, through prayer, they ended up on the next flight out. We departed and headed to Rome for a short lay-over and then on to the fifteen and a half hour flight to Washington D.C. and then to Ohio. When we arrived at Dulles Airport in Washington D.C., it was somewhat congested, and we really didn't get a chance to say proper goodbyes to many of our newly found friends. But through the use of email and cell phones, we would eventually make contact with many of them again. Our team included people from Georgia, Iowa, California, Colorado, and Peru. It was a great experience for all, and I pray for them often to "Arise and go" with what God shared with them on their journey.

Within a few days after returning home, I had recovered from the jet lag and began further research, starting in Ethiopia and then from Bob's findings. I buried my head into manuscripts, church historical documents, and historical writings of Jesus' trek into Egypt. Shortly after our return, phone calls started to come in from organizations and churches that wanted to know about the journey. As a result of this interest, I hit the circuit sharing what we had discovered about the Ark and our newly found passion for these people of Ethiopia, while keeping the knowledge of Christ's appearance on this journey to a minimum and continuing my research. My wife was hired by a home improvement store which was a tremendous blessing to help offset travel and living expenses, while I continued the operation of our company, Cross Point Magazine.

TRANSLATING THE BOOK

One Wednesday afternoon in May 2011, Dave Kochis called me from San Francisco to discuss our plans for the return trip to Ethiopia. We strongly agreed we needed to have Bibles printed in the Amharic language to take back to the people of Ethiopia, and he wanted to know if we had any success in that translation yet. I shared what had just happened earlier in the week while attending to a funeral of a family who had lost a young child. Across the parking lot from the church was a Bible publishing company that I thought may have this translation. They serve numerous missionaries, so I checked with them before the services to see if they happened to have the Amharic translation of Scripture. Before going there I knew that the chances were very slim that they would have this particular language because of its secluded use in Eastern Africa. This particular publishing company serves missionaries with Scripture all around the world, but I knew there were none in the Amharic Ethiopian region, but I thought it would be worth the try anyway.

I stopped in and caught the director of the publishing house getting ready to leave for lunch. He was kind enough to take the time to answer a few questions and to discuss the language. He pulled out a book and flipped through it and then turned around to his computer and said, "Let me look at another place on here." My anticipation rose as he tried to locate the language in question. Seconds seemed like minutes while his fingers continued to scroll down the pages. Finally, he turned and said, "I really don't think we've done anything like this before. I wish I could help, but I'm afraid that we don't have this translation." I was deflated. It was a flicker of light and then a breeze blew it all out. I thanked him for his time and told him I had to go to a funeral at the church. Somewhat disheartened, I hopped into my car, drove to the church, and went into the service realizing I needed to perk up to support the family. I was sitting in the back of the church during the service when a man tapped my shoulder, startling me from my thoughts. He leaned over and handed me a note. It read, "Al needs to see you over at the print shop." I looked at him and nodded blankly.

As the service wrapped up, I paid my respects to the family and headed out the front door. I felt like a NASCAR driver as I spun-out trying to return to the print shop quickly. I pulled in by their entryway, literally creating my own parking space and headed back into the publishing company. The director was waiting for and waived me down the hallway to his office. I walked in; he stood holding a small green paper covered Scripture and a floppy disk. Yes, an old floppy disk! He said, "I don't know when we would have even done this because we don't have missionaries needing this translation, but I found it." I said, "You found what?" He replied stuttered along, "It's the Am, Amhara, no, Amharic language.

We have it. I don't ever remember having this translation, but this is it on this old floppy. I found with it a translation of John and Romans as an example. I was surprised but excited to tell you. So we can do this for you." It was like a disbelief relief. I grabbed the book from his hand and started flipping through the Scripture. Not that I could read it, but it was exciting to see. I shook his hand and said I would get a date to plan to bring a team in to print and prepare the Bibles. He agreed, we shook hands again and I moved quickly down the hallway. It was more like a skip down the hallway, my heavy heart lightened. Finally, one phase of my search and return to this ancient land was complete, for the most part.

Dave was excited that I was able to find the translation for the Bibles. He then inquired about the translation of the Bible pages from Ethiopia on Tana Kirkos Island, but I had no luck on that yet. One road after another would lead to a dead end. I just didn't want to turn the pages over to someone who had no respect for what we were trying to accomplish. These were very holy documents and had been under lock and key for thousands of years. I had been entrusted with them and felt somewhat like a protector of the story until time to share it with the world. He then closed our conversation by saying that he was going to pray that we could find someone to translate this ancient language of Ge'ez from the book we had photographed on the island. He said, "Jim, do you realize how important this information is to understanding Jesus' life as a boy and His future ministry? This has been hidden for centuries and this portion is completely unknown, if that's what it says and you can confirm it." I simply said, "I know." We ended our conversation with anxious goodbyes.

The next day I was in South Central Ohio when Sherri called me, "Jim, you have to come down here, there's a woman that started working here and she is from Addis Ababa, Ethiopia. She says she might be able to help us with the translation of the book." In disbelief, I told her I was on my way. It was about a two hour drive to where my wife worked. I knew this could be my last shot to find someone who knew this ancient, extinct language, rather than waiting to return to Ethiopia. I waited nervously in my car until Misrak, the translator was off work. As she approached, I noticed her quickly. She had that Ethiopian smoothness to her face and very distinctive appearance. We introduced ourselves and she said, "Let me see what you have so I can tell you what it says." I handed her the scans of the pages and she suddenly stopped. She said, "Oh...I no read this. This is from Ge'ez, you know, and only a priest can read this. No one speaks this language and only the priests can tell you what it says." I was shaken, and it seemed my last hope for a translation had been dashed until returning to Ethiopia. Then, quite unexpectedly she added, "I think I might know someone who can read this." I said, "Who?" She quickly replied, "My priest at Ethiopian Church in Cincinnati." In shock I stood there having no idea that an Ethiopian Church existed in Cincinnati. When she promised to ask him, I felt like hugging her but that would not have been inappropriate. Instead I grabbed her hand to shake it vigorously and thanked her several times for taking the time with me.

I called Dave in San Francisco and told him the anticipated good news and he was elated. I belted out, "God answers our prayers when He has a mission to glorify Himself in the process, doesn't He?" Without hesitation, Dave replied, "Always. I'm so excited about this. Please keep me informed." It was about a week later when Sherri called and said she had priest's number. I called his home, but he spoke no English. One of his children picked up the phone, and although the English was much better, it was still too difficult to understand on the line. The only thing I could make out was, "We will call you." I didn't know if that was good or bad. I began to think the latter as days had gone by and I heard nothing. Then I called back and received the same answer from a woman with broken English, "We will call you" CLICK!

Three weeks had now passed into mid June and I was growing more anxious, but at the same time almost hopeless. Then, again on a Wednesday morning around 8:00AM my phone rang. But when I answered no one was there. I figured if it was something important they would leave a message or call back since I had no idea who this caller was. Call back he did in thirty seconds. The man uttered in broken language, "Hello, is this Jim?" I answered expectantly as I recognized the dialect, "We will meet you today at main library, Cincinnati at 5:00PM. Bring the pages with you...goodbye." Conversation was over! I was caught in the middle of clandestine rendezvous to decipher clues about some hidden treasure. The anticipation mounted, and I called Sherri to tell her about our 5:00PM appointment.

The afternoon passed slowly by as I organized my thoughts and questions. Finally, three o'clock arrived, and we headed to Cincinnati. We first purchased a new digital voice recorder at a nearby electronics store. It was time to retire my old unreliable cassette recorder. I didn't want to risk missing a syllable during this mysterious meeting. We drove around the streets in downtown Cincinnati looking for a parking place near the library, which sprawls out over two sides of the streets, connected by a walkway over the busy intersection. I found an empty metered space around the corner, parked, and went in. I couldn't sit down, so we looked around and found Bob Cornuke's book, *Relic Quest* in their wide expanse of reading. Five o'clock drew near and I realized that sometimes these priests are uncomfortable with women present. Sherri agreed, remembering the required dawning of prayer shawls necessary while in exclusive areas of extreme holiness that women were not welcomed in Ethiopia. She wandered off while I stood pacing and profiling each person who entered the three entrances of the massive library. I noticed a small older man with a white beard carrying a satchel standing by the entry to the audio section; he looked Ethiopian and appeared at peace with the world, but he was alone and the voice earlier plainly said "We will meet you," so I knew there had to be someone else nearby if that was him. At that moment my phone rang, "Jim, are you here?" came the voice both in the phone and behind me. The other much younger man with an unmistakable Ethiopian look was standing next to me. I turned in surprise and said, "It's me...I'm Jim." He then motioned to the Abba, the white bearded older man to join us.

After warm handshakes and introductions, we made our way to a back corner table to be alone. The priest, Abba, took the scanned pages and began to read in a quiet voice. Yonas, the younger man, and I continued in conversation, and I expressed my thankfulness for meeting with me. He replied with the familiar simple normal Ethiopian response "No problem." Yonas was from Addis Ababa, and the Abba was from all over northern Ethiopia. Twenty minutes passed by before the Abba broke his stare from the pages, still reading under his breath. Then he turned to Yonas spoke in the more modern Amharic. "The Abba says, this page (reading from the left column of the all writings page) says this is about a meeting of like churches to settle differences and come to Egypt. God will bless them for that. This is written in a book in Alexandria library. The rest of this is not." I was momentarily disheartened as I was hoping it would contain information about Jesus in Ethiopia, when he said, "The rest of this is forever Ethiopian, like secret. No one knows the rest of this." Now they had my full attention. "Does it say anything about Jesus going to Ethiopia?" I asked. He turned to the Abba and asked that question. The Abba slipped his glasses down and started to read again. It was like he wanted to make sure that what he was reading was, in fact, what he thought he was reading.

THE CONFIRMATION: JESUS IN ETHIOPIA

Abba Yemanebirhan looked up and spoke to Yonas, who turned to me again and said, "This writing is something no one else knows." He referred to the right side of the first page from the book, again showing nothing but written script in Ge'ez. He said that it talks about Jesus, Mary, Joseph and another woman known in the Scriptures as Salome. Do you know Salome?" he asked. The Abba began to speak again as I nodded in recollection of this woman. Salome was the wife of Zebedee and the daughter of Mary's mother's sister, also named Mary, referred to in the Scriptures as "the other Mary" *(ref. 9)*. Salome is mentioned in the Bible several times including her appearance at the cross with Mary:

"There were also women looking on afar off: among whom was Mary Magdalene, and Mary the mother of James the less and of Joses, and Salome;" Mark 15:40

And then it goes on to describe who her importance in verse 41:

"Who also, when he was in Galilee, followed him, and ministered unto him." Mark 15:41

Then again:

"And when the sabbath was past, Mary Magdalene, and Mary the mother of James, and Salome, had bought sweet spices, that they might come and anoint him." Mark 16:1

This is completely new-found information that Salome was with Holy family during this trek. This may explain her obedience to Christ throughout the Scriptures and even until His crucifixion on the cross. She was definitely a part of his life since he was born because of her relationship to Mary. This young mother may have asked Salome to join them on their journey to help with her Son during this strenuous trip through Egypt. Regardless the reason, this is new information which plays an important role in understanding the complete historical account of this journey.

Yonas turned to me and said, "All of them traveled into Egypt after Joseph was visited by the angel, you know, that told them to leave Israel to escape Herod's killing of the children." The Abba continued to speak, and Yonas continued to translate. The information was plentiful. He then said something that would really catch my attention. "This book was written by Saint John; the Apostle John was told this from Saint Mary and written down by student of Saint John. She told him about it later" he said. Excitedly I belted out, "John wrote this story from Mary's recollection?" Yonas calmly replied, "Yes, Saint John wrote this and he tells of them going to Egypt to escape from Herod killing all the children. It says here 8,000 children, but may be more because the page is messed up." What's really overwhelming, John is the son of Zebedee and Salome. Yes, the same Salome that made this journey with Jesus, Mary and Joseph is also the mother of the man who transcribed the story from Mary's recollection. Could Salome have been the initiator that brought Mary to John in order to give her this incredible story after Christ's ascension into Heaven? It seems as though this woman, Salome, played an important role from the time of Jesus' birth to realizing the importance of recording the story of the miraculous meeting for future generations to hear.

Yonas continued by sharing another bit of awakening information in the book, "It says they traveled through Egypt with help of the archangel Uriel." I continued, "Is this the angel that came to Joseph to tell him to flee?" "I don't know" he said, "It doesn't tell us here." Uriel was one of the four archangels (Michael, Gabriel, Uriel and Raphael) whose name literally means the "Fire of God" or the "Light of God" *(ref.11)*. As Jesus is looked at as The Light, it makes sense that the reference to an angel guiding them with a name meaning "The Light of God" would come into play.

Yonas continued to translate from the Abba's readings; "It says here that Saint Mary told Saint John about their trip in Egypt and down to Ethiopia. This part here (pointing his finger at the etched drawing of the Holy family with the angel) tells about them coming into Ethiopia; Mary, her Son Jesus, Joseph and the other lady, Salome with the angel Uriel. This says that they went together to Lake Tana in Ethiopia and this is the church" (pointing to a picture etched on the page of drawings). I cautiously sought confirmation, "Where is this church?" He said, "The church in Lake Tana." I shocked them with my next question, "Is this the tabernacle that was on Tana Kirkos Island where the Ark of the Covenant was?" Both Yonas and the Abba turned and looked at each other, and then Yonas said, "You know Tana Kirkos?" "Yes, I've been there and that's where I took these pictures of the book from." The Abba raised his head in agreement, stroked his beard, and said something to Yonas. He turned around to me and said, "Then you know that this picture is Maria, Jesus, Joseph, Salome, the angel Uriel in a boat going to Tana Kirkos to be with God?" My heart quickened as I replied, "I wasn't sure, but I was hoping that was what you were going to tell me." Yonas continued, "This is the church (he paused and listened to the Abba for a moment), the place where Jesus would go to pray with God." I came back with the question, "Do you mean the tabernacle where the Ark of the Covenant was at?" Jonas nodded and then answered, "Yes, the church is like a tent or yes, the tabernacle on Tana Kirkos." The tone of the conversation began to change a little, and I wasn't sure whether it was a good or a bad change at first. But it seemed as though they began to talk with me a little more freely. Yonas stated, "This has been a secret in Ethiopia for two thousand years and few people know about this, about Jesus and Maria coming to this place." I nodded to confirm that I completely understood. It was now more evident than at any other point that God had maneuvered information into my hands for a reason. I believe that these men were sent to me to confirm that all that I had prayed about and experienced was recorded on these pages.

Yonas asked about the next page of the book which consisted of only eight and half lines. He asked me if this was in the beginning or end of the book and I told him that it followed these pages. He shook his head to confirm that I was telling him the truth-that this was supposed to follow the other pages. He explained, "When Saint John wrote this down from Mary telling him about the travel here, he is telling anyone who reads this not to take out, put in, or change anything in the book because God will not be with them. He will not bless them." I said, "We have this written by John in the book of Revelation." He agreed, "Yes. Yes. This says not to change the words or do anything to change what it means, just like Revelation, but this was written first before that in Revelation." I then went back to the drawing of Mary, Jesus, Salome, Uriel and presumably Joseph sitting on a chair and asked what exactly they were doing and what this writing said. I had forgotten what Misgana had told me on the cliff, next to the site where Mary had watched Jesus pray at the tabernacle, that Joseph was not on the island with them. He had told me that Joseph was working as a farmer on the mainland to earn money for their travels. That led to the assumption that Mary, Jesus and Salome were on the island and taken care of by the monks and priests who stood guard over the island. Yonas turned to the Abba, and they conversed for a few minutes. Then Yonas said, "This is God holding the law and teaching Jesus for later. He is at the church, or the tabernacle like you say, on the seat teaching Jesus all He will need." "The seat" I asked? "Yes, the seat, (he paused to gather words) the Ark where God would come." I was speechless. I actually had to gather my thoughts and force myself to breathe again. Everything was coming to light, and I really had no more questions. This was beyond everything I had hoped. Everything was now confirmed, down to the last detail. Everything I had been told and everything I had experienced on the island was verified by these two incredible gentlemen.

Sitting at the table, trying to absorb this revealing information, after nearly an hour and a half of questions, the Abba seemed to be getting tired reading the pages. However, he told me about the last page of the photo scan that talked about the fallen angels before Creation and how ninety-nine angels had driven them out of Heaven. Then, the Abba, scooted back in his chair, as if to say "we are done." I stood up and had to ask one more question. Well, it actually turned into two questions. First I asked, "Do you believe that the Ark of the Covenant is in Axum?" The Abba, understanding most spoken English but unable to respond in English, nodded while Yonas answered, "Of course, it is!" To question that truth suddenly seemed silly. My second question, "Do you really believe that Jesus and Mary came to Ethiopia?" Without hesitation he answered, "There is no question they were in Ethiopia. This Bible tells us that He was there and the tradition is true. You see it here with your eyes." The Abba talked with Yonas again, and he turned to me with even more revelation. "When they came to Ethiopia they went to Axum and then taken by the angel to Lake Tana, Tana Kirkos." I asked why they would have gone to those places. He said that's where they went when they came to Ethiopia to bless that place. "When on Lake Tana it was very important for them. They went to Tana Kirkos with God," he added. "This book was brought to Ethiopia, to Tana Kirkos by John's student, I think you say Prochorus. It is important secret of Tana Kirkos. This is the place where Jesus and Saint Mary go to be with His Father."

Prochorus is noted in the Scriptures as being one of the seven deacons that were hand- picked to take care of the poor and to continue to spread the Word of God.

"Wherefore, brethren, look ye out among you seven men of honest report, full of the Holy Ghost and wisdom, whom we may appoint over this business. But we will give ourselves

continually to prayer, and to the ministry of the word. And the saying pleased the whole multitude: and they chose Stephen, a man full of faith and of the Holy Ghost, and Philip, and Prochorus, and Nicanor, and Timon, and Parmenas, and Nicolas a proselyte of Antioch: Whom they set before the apostles: and when they had prayed, they laid their hands on them." Acts 6:3-5

Prochorus is said to have been the nephew of Stephen and traveled with the Apostle Peter during his early ministry *(ref. 22)*. Sometime later, it has been recorded that he was a student of John and became a scribe for him. John ordained him to become the bishop of the church in Nicomedia and it is recorded that Prochorus was banished to Patmos with John. It is also believed that Prochorus may have been the man responsible for scribing John's vision for the Book of Revelation *(ref. 23)*. He eventually left Patmos and continued to share the gospel of Christ before suffering the death of a martyr. This revealing information of who brought this information to Ethiopia fills in another piece to the puzzle, as Prochorus being the connection to the same warning from John in Revelation 22 and in the eight and a half lines in this writing. Then John's connection to Salome, as his mother, also adds to the equation of her bringing Mary to him to write her miraculous story about their journey to Tana Kirkos. Yonas wanted to make it clear that John wrote this book and Prochorus was responsible for bringing it to Ethiopia and to Tana.

Yonas pleasantly extended his hand to shake mine, with his other hand gripping his forearm in the traditional Ethiopian handshake. I did the same in affirmation with their custom. The same was done with the Abba, along with a slight bow to acknowledge the mutual respect we have for one another. We continued our conversation walking through the building to the front door of the library where Sherri stood waiting anxiously to hear the result of our meeting. We talked for awhile as Yonas invited us to their Ethiopian church in downtown Cincinnati for an upcoming celebration a little over month away. We again shook hands and parted ways. I was truly exhausted but thrilled with the overwhelming confirmation wrought during this secret rendezvous in the city. It was a confirmation that brought life to pages of the book with the Ark of the Covenant, God's presence, Salome, Prochorus and the Apostle John once again as the Revelator. But most of all, we see that this small lost island was the place where Jesus and His family traveled to gain knowledge from His Father, God Himself. It's a perfect island in many ways that has been kept from the sin of the world since the beginning of God's Creation. This was also a confirmation that God had cleared this path for me to follow, and I was not about to lay it down and let it pass by now. It confirmed that for me it was time to "Arise and go!"

Chapter 12

SO...WHY ETHIOPIA?

Many people have offered opinions about the importance of Ethiopia to biblical history. There is no denying that Ethiopia is one of the most mentioned countries throughout the Scriptures. They play enough importance that God included them in key Scriptures and some of the most important roles in history, both in the past and in the future. From Moses' wife Zipporah, to Solomon's attraction, to the return of the Ark of the Covenant, to the reasons why Luke spent so much time on the Ethiopians in Acts chapter eight, Ethiopia's masterful role in history proves unquestionable. One thing is for sure, this African country may hold more than the secrets of Jesus' past, it may also hold the keys to Jesus' future reign.

River Near Eden

One of the most interesting Scriptures in the early texts of the Bible came with the foundation of the earth in Genesis:

"And a river went out of Eden to water the garden; and from thence it was parted, and became into four heads...And the name of the second river is Gihon: the same is it that compasseth the whole land of Ethiopia." Genesis 2:10, 13

This particular Scripture has generated much controversy about the exact location of the Garden of Eden. Evidently, Ethiopia was greatly treasured by God to include them in the opening pages of His Holy Word.

Moses' Ethiopian wife:

Another clear clue to the importance of Ethiopia in the Scriptures comes with the marriage of Moses to an Ethiopian woman of color:

"And Miriam and Aaron spake against Moses because of the Ethiopian woman whom he had married: for he had married an Ethiopian woman" Numbers 12:1

We see in this Scripture the disagreement about Moses' marriage to a dark skinned woman, more than likely named Zipporah (Exodus 2:21). Because of this disagreement between Moses and his siblings, Miriam (who appears to be the instigator since she is mentioned first) and Aaron, God cursed Miriam with leprosy to show His anger upon for her apparent racial judgment. Of course, after much crying out in prayer by Moses, Miriam was healed of her curse.

What was the real reason why we were brought to the knowledge of Moses' wife? Could this mention in Numbers be our introduction to possibly foretelling of the resting place of the Ark of the Covenant to be guarded by the Ethiopians until the time it is to be returned to Jerusalem? Could it also be a telling account that this is where Jesus, the Messiah, would travel to in order to meet with His Father for a quarter of a year to gain the knowledge and blessings for His future? To me it is interesting that God would have chosen certain passages to even be included in Bible text, if they didn't have a further meaning. We have to assume that the passages mentioned here were to give us a closer look into the future reference to Ethiopia and its importance to the times to come, referring to the Mercy Seat, the Ark of the Covenant and the protectors of these artifacts. Everything included in the Holy

Word of God has an importance in the past, present and the future. Again, this appears to have a more intricate role than we may have first thought.

The Prophetic Scriptures of Isaiah

Even though there are many other Scriptures that precede Isaiah 18 that mention Ethiopia, this is one of the most prophetic Scriptures that alludes to the whereabouts of the Ark of the Covenant's current location and future events leading up to the Ark's return to Jerusalem. On the cliff side overlooking Axum, Ethiopia during a cool evening, Bob Cornuke and I were relaxing in our chairs next to small fire that one of the motel workers had started for us after our long day of work in a makeshift clinic and exploration of the tombs of the kings. Even though I was wearing a jacket because the warm days had turned to star studded cool nights, I can remember that I couldn't wait to pop open an ice cold Pepsi to splash some taste of home on my palate and wash down that gritty taste of dust and sand. As Bob and I sat talking about the next day's adventures, he asked if I had read his book *Relic Quest*. I told him that I honestly had only read the parts that had nothing to do with the Ark of the Covenant because I didn't want a preconceived bias about my purpose on this exploration. Bob then asked, "Have you read Isaiah 18 yet?" Without any hesitation I told him that I had read Isaiah many times before, and I considered it one of my favorite and most intriguing books of the Bible. He nodded his head and the conversation shifted as Dave Kochis and Craig Newberry pulled up chairs along side of us.

This topic never came up again until our team sat on bench seats in Barhir Dar one evening for a Bible study just a couple of hundred yards from the shore line of Lake Tana. Beneath the lobby canopy of the Kariftu Resort, which looked much like a thatched hut roof with no doors and an open entry right to the front desk, the team gathered to listen to Bob's explanation about the strong possibility that the Ark could still be in Ethiopia today. For the next hour and a half I had a flurry of thoughts, ideas, and revelations spinning in my head. Pieces of our journey that had already taken place or ideas impressed upon me started to make sense and all began to fall into place.

Even though this story is not about the Ark of the Covenant specifically, its appearance is crucial as the place where God came to meet with His Son. With this evidence as truth, we must be able to place the Ark somewhat in this region of the country for this particular encounter. During the study,

Cornuke opened up his Bible and started to read from Isaiah 18, and it was like all of the events beginning in the middle of the night in Gatlinburg, Tennessee to kneeling in front of the trumpets in Axum had now become a bright shining light of reality to me. It was like a look into the future in real time and place unveiling before my eyes. As my thoughts began to wander, I hurriedly scratched notes and flipped open to other passages where I had made notes along this journey. Isaiah 18 elevated this entire journey from an "everybody has their own thoughts" to "Oh, my goodness, have we just walked on Holy ground" reality. Let's look at this Scripture closely:

"Woe to the land shadowing with wings, which is beyond the rivers of Ethiopia: That sendeth ambassadors by the sea, even in vessels of bulrushes upon the waters, saying, Go, ye swift messengers, to a nation scattered and peeled, to a people terrible from their beginning hitherto; a nation meted out and trodden down, whose land the rivers have spoiled! All ye inhabitants of the world, and dwellers on the earth, see ye, when he lifteth up an ensign on the mountains; and when he bloweth a trumpet, hear ye. In that time shall the present be brought unto the LORD of hosts of a people scattered

and peeled, and from a people terrible from their beginning hitherto; a nation meted out and trodden under foot, whose land the rivers have spoiled, to the place of the name of the LORD of hosts, the mount Zion" Isaiah 18:1-3, 7

Breaking this down more closely, we see in verse one that the thoughts on this passage describe Ethiopia as it is now, not necessarily as it was when Isaiah wrote this Scripture. Some say the *"shadowing with wings"* could be the flies buzzing around or the many eagles that soar through the skies over the region. It's followed by a passage pinpointing the region *"beyond the rivers"* of Ethiopia which flow greatly in the northern mountain regions of this country. It goes on to tell us about ambassadors who will be traveling in the reed, bulrush, and boats that the fishermen of Lake Tana still travel in today. The Scripture continues to describe the people of Ethiopia and how their nation has been, somewhat sacrificed in order to protect something of great significance and value. This includes the disease and even the spoiled waters that they are plagued with today. But then it turns to a time when there will be a blast of the trumpets from the mountains of Ethiopia. The only region, of Ethiopia with mountains is the northern region and it makes you wonder if the trumpets could once again be those housed today in the treasury of the Ark in Axum. These trumpets, the trumpets of Moses, were blown to call the people to the tabernacle for instruction (Numbers 10:1) as well as for the announcement of the Ark's arrival when it was moved from place to place;

"And David danced before the LORD with all his might; and David was girded with a linen ephod. So David and all the house of Israel brought up the ark of the LORD with shouting, and with the sound of the trumpet. And as the ark of the LORD came into the city of David..." 2 Samuel 6:14-16

Finally, a sign will be brought forth in those mountains and verse seven shares that a singular present, or gift, will be brought from Ethiopia to Jesus in Jerusalem and it will be taken to Him in what it describes as the place where the Lord of Lords will rule from in the Holy of Holies in the third temple. When we re-examined this Scripture, it stopped me cold. I asked Bob what this specifically meant because for years, I had incorrectly interpreted it as meaning the "present time" instead of a present, or gift, brought in an appointed time period. Bob shared that this verse had also been a stopping point for him during his many years of research on the

Ark and its possible resting place in Axum today. No matter what, we know that something of great value to Christ will be revealed from the mountains of Ethiopia and will go to Him in the *"place of the name of the LORD of hosts, the mount Zion"* in which we know from many other Scriptures is the Holy of Holies, where Jesus will take up His throne. With even closer review of Scriptures, we know that the Mercy Seat, the solid gold covering of the Ark of the Covenant, was actually the throne of God's appearance on this earth as He would *"dwell between the two cherubim."* One could speculate through the many Scriptures that support the fact, that this same Mercy Seat will one day return to Christ as His throne and He will reign from that same place.

Without any doubt, this Scripture is a compelling glance into a future event with Isaiah pin-pointing Ethiopia as we know it, not the one with a building empire found during that particular time in history. To set up the Ark's movement from Jerusalem, Cornuke shared that at the same time as the evil King Manasseh's reign and defilement of the first temple, another temple matching that of Solomon's dimensions had been built in Aswan, Egypt on the Nile River of the island of Elephantine. This southern Egyptian location can be confirmed through many Levitical papyri and scrolls left at the site. And with an incredible view of the Ark's presence with the Egyptian Pharaoh, Necho in 2 Chronicles 35 taking his orders to march into battle directly from God Himself, we have great evidence that the Ark was moving to the south. With a pagan god-driven nation now getting its orders from the one true God, we can assume the faith of this particular pharaoh and his accepted belief in Him.

Bob then explained that the Levites, according to historical records, simply left the site of the Elephantine Island temple and moved on to somewhere else. It was within a few years that the Ark's arrival was reported on the island of Tana Kirkos which would have been a straight journey to the south down to the Nile to the Blue Nile and into Lake Tana, its feeding point. From here, the Ark rested on this island in a makeshift tabernacle for nearly 800 years until King Ezana claimed it to his kingdom in Axum. The most amazing aspect of these claims is the overwhelming evidence that backs each step of this journey. This is a journey that many have overlooked or dismissed over the years. One reason I believe it is overlooked is the power of tradition to shy away from anything that may be new information that could challenge what man has decided is truth and not what God has shown us as proof.

The Controversy of Jeremiah

After I had began the speaking circuit after my journeys into Ethiopia, the biggest question that always came up was the use of Jeremiah 3:16 as the Scripture that was to prove that we don't need the Ark of the Covenant and that it is gone forever. I loved it every time I would hear folks discussing this Scripture before I ever took the podium. Once again and rightfully so with the standard tradition, we see what appears to be a prophecy of Jeremiah exclaiming that the Ark no longer is needed in history:

"And it shall come to pass, when ye be multiplied and increased in the land, in those days, saith the LORD, they shall say no more, The ark of the covenant of the LORD: neither shall it come to mind: neither shall they remember it; neither shall they visit it; neither shall that be done any more." Jeremiah 3:16

This verse does tell us that there really won't be any need for the Ark in history again. It has served its purpose as Jesus came to us in order to become the Ark while holding the Law of God within Him. What many traditionalists don't do is read on to verse seventeen which gives the rest of the story:

"At that time they shall call Jerusalem the throne of the LORD; and all the nations shall be gathered unto it, to the name of the LORD, to Jerusalem: neither shall they walk any more after the imagination of their evil heart." Jeremiah 3:17

Completing the story, we see a period of time will pass once again. It appears as though this is the same time period that Isaiah mentioned in verse seven of chapter eighteen proclaiming that there will be a specified time period that Christ will take up His throne in Jerusalem in the Holy of Holies in the Temple. Again, we can gather from numerous other biblical texts that the mentioned throne will be the Mercy Seat which rests above the Ark of the Covenant. As the Ark itself will no longer be needed, we also see that the separate covering of gold, the Mercy Seat, will play a most intricate role upon the return of Christ.

Ezekiel's Look into the Throne Room

During our study in the thatched foyer in Barhir Dar, Bob began to bring forth numerous connecting Scriptures to support Isaiah 18 about the location of the throne, as well as what that "present," or gift, will be. It quickly became evident that we were dealing with something at the time beyond our traditional understanding but one that demanded more Scripture to confirm his comments.

One of those Scriptures came in the prophecy of Ezekiel:

"And the glory of the LORD came into the house by the way of the gate whose prospect is toward the east. So the spirit took me up, and brought me into the inner court; and, behold, the glory of the LORD filled the house. And I heard him speaking unto me out of the house; and the man stood by me. And he said unto me, Son of man, the place of my throne, and the place of the soles of my feet, where I will dwell in the midst of the children of Israel for ever, and my holy name, shall the house of Israel no more defile, neither they, nor their kings, by their whoredom, nor by the carcases of their kings in their high places." Ezekiel 43:4-7

These verses reveal a prophetic place upon Christ's return to take over the third temple in Jerusalem, and more precisely, the inner court of the Holy of Holies where His throne will be placed. Never in the history of any king did they reign on the throne from inside the temple. They always took their place of rule from their own thrones within their own palace or castle. This verse specifically is talking about the throne, or the Mercy Seat, being placed inside the temple upon Jesus' return, and His ruling from the Holy of Holies on that throne where He will place the souls of His feet and reign from there forever. We know of the time period, upon His return to claim the throne forever, because Ezekiel describes the time when there will be no more defilement or sin during this reign.

This verse brings to light the days that are to come when Jesus makes His appearance and what His Throne will appear as. Placing this next to Isaiah 18 brings us closer to piecing together those events that are yet to come. Could this have been part of what God revealed to Him on Tana Kirkos? We know that God rested on the Mercy Seat, His throne, on the island and taught Jesus from there during the three months and ten days of prayer and teaching. It leads to speculation on the overall knowledge that was given to Jesus at this meeting and to what extent He was preparing when He left.

Zephaniah and the Royal Procession

By this time of the evening, Bob was obviously growing tired from nearly a month of non-stop travel before joining up with us on this expedition. You could tell he was wearing down, but you also could see his thrill to share more with us. It was as if he were rediscovering what he had uncovered in previous times. Regardless of his tiredness, he continued to make sure we had complete knowledge of the Scripture to support his theory on the Ark and its relevance and relationship to Ethiopia.

He continued to look into the Scripture from Zephaniah which once again brings to the forefront a special offering from the people of Ethiopia:

"From beyond the rivers of Ethiopia my suppliants, even the daughter of my dispersed, shall bring mine offering" Zephaniah 3:10

This Scripture moved us all. Earlier in the day we had taken the long journey through the desert from Gondar to Barhir Dar. Before leaving Gondar we had the honor of meeting the lady who is one of the remaining Falashian Jews in Ethiopia. When reading the Scriptures in Zephaniah, one can't help but wonder if a descendent of Marye from Gondar could have a direct hand in bringing forth the Ark back to Jerusalem when it refers to *"the daughter of my dis-*

persed." But what's even more compelling is the way the Ark and Mercy Seat will be returned. If you separate the word "bring" in this text to a translational context, you will start to see a procession opened up, much the same as you would see in Axum during the Timkat ceremony. Could this ceremony possibly even be a yearly rehearsal, if you will, to the events to come somewhere in the future?

"Bring" in this text is taken from the Hebrew word "jawbal" which means to "bring forth" or to "lead forth" in a ceremonial type procession with pomp and circumstance. Sitting under a thatched roof in the middle of Ethiopia, along the banks of a lake that claims to have housed the Ark for eight hundred years, it seemed as though things were really "clicking" in my brain at this point. I realized I may have witnessed a yearly practice of the "when it's time" return of the throne of Christ from the people of Ethiopia to the King of Kings in the house of the Lord in Jerusalem.

Had these people always recognized this ceremony or were we the ones who have never picked up on this procession. During the Eve of Timkat, which is again claimed to be the celebration of the baptism of Jesus (among other speculations), the monks and priests carry out a replica ark, or Tabot, covered in a jeweled covering. They parade it from the Saint Mary of Zion Church through the streets to a makeshift tabernacle where they place it in the Holy of Holies of the tabernacle. They do this dressed in the ceremonial Levite style robes and gowns while blowing trumpets, pounding kebero drums, and clanging sistras. We know from Isaiah 18, "*All ye inhabitants of the world, and dwellers on the earth, see ye, when he lifteth up an ensign on the mountains; and when he bloweth a trumpet, hear ye*" which states that when the "*present*" leaves Ethiopia, that a sign will be shown in the mountains and all the world will hear the trumpets blow. Could this be a secretive rehearsal for this day? The thoughts kept running through my head. Later I asked Sherri, Dave, Anita, and Steve and they all were of the same feeling. I concluded that while we were in Axum, we should have paid closer attention to what we were seeing because we may have missed something vital to our research.

Nevertheless, we were enlightened by the Scriptures according to our understanding. They were beginning to unravel the mystery of the sign, the present of the returning throne, and the procession that will celebrate the Mercy Seat's travel to the inner court of the third temple in Jerusalem... "*place of my throne, and the place of the soles of my feet.*"

The Scripture to back these theories is overwhelming as we found in various other places including Psalm 68:29, *"Because of thy temple at Jerusalem shall kings bring presents unto thee."* Once again the word "jaw-bal" comes into play with the word "bring." It's somewhat mind blowing, but after returning to the United States I quickly found that this would all precede further evidence to qualify the truth between God the Father and God the Son on the island. It seems as though the entire journey may have been a topic on the island for all of the future events still to come.

Reigning from the throne

More evidence of Scripture foretells the building of the temple and the overall presence of Christ taking up His throne there:

"And speak unto him, saying, Thus speaketh the LORD of hosts, saying, Behold the man whose name is The BRANCH; and he shall grow up out of his place, and he shall build the temple of the LORD: Even he shall build the temple of the LORD; and he shall bear the glory, and shall sit and rule upon his throne; and he shall be a priest upon his throne:

and the counsel of peace shall be between them both." Zechariah 6:12, 13

There is no doubt that when that day arrives there will be no question that Jesus has taken rule and from where He will reign. Scripture is clear several times over staking this claim. My mind returned to Axum once again. One day while in this humble place, we were granted the honor of receiving a blessing from Guardian of the Ark, the man who gives his life inside the enclosure of the Saint Mary of Zion Church to guard it, and another blessing later in the day by Narub, the high priest of Axum. When we quietly met with the Guardian, he inside the iron fence of the church and we outside, that we were less than twenty yards from the corner of the building where the chamber that contains the Ark of the Covenant is supposedly housed.

The possibility that the throne which God sat upon, the Mercy Seat that He met with His Son upon at Tana Kirkos, the place between the two cherubim, and what could be the future throne of Jesus Himself in Jerusalem, might just have been a few yards from me in this small building. This thought literally brought shivers up my spine. Could I have been wrong all the years? Had tradition clouded me so strongly that I was unable to see what God was showing me so plainly? From this point on, I decided that my mind, my heart and my trust were going into whatever God had in store for me. After that night of Bible study, I was thankful I made that commitment, because it was the next day that we ventured to the island of Tana Kirkos and my true adventure really began.

Ethiopian Eunuch

This is another wonderful example of a biblical story, or event, presented in great detail, but with the true purpose possibly overlooked. When Luke described the events of Phillip and his ministry in Luke 8, he gave a tremendous amount of time to one particular person in this passage. The Ethiopian Eunuch has been the example for many pastors over the ages as the perfect standard for salvation and the immersed water baptism. What could possibly be overlooked is Luke's apparent clue to the importance of this man and the story's relevance to the Ark of the Covenant's location.

Whenever I come across a biblical event that gives great detail, I become inquisitive why God allowed this much information to be given. This is what Bob Cornuke brought out in our study in Barhir Dar under the canopy of the motel.

I had always pondered this detail but had fallen into the accepted tradition of what this Scripture was intended to be used for. That, of course, was to explain salvation and baptism to a possible new believer in Christ.

"And the angel of the Lord spake unto Philip, saying, Arise, and go toward the south unto the way that goeth down from Jerusalem unto Gaza, which is desert. And he arose and went: and, behold, a man of Ethiopia, an eunuch of great authority under Candace queen of the Ethiopians, who had the charge of all her treasure, and had come to Jerusalem for to worship, Was returning, and sitting in his chariot read Esaias the prophet. Then the Spirit said unto Philip, Go near, and join thyself to this chariot. And Philip ran thither to him, and heard him read the prophet Esaias, and said, Understandest thou what thou readest? And he said, How can I, except some man should guide me? And he desired Philip that he would come up and sit with him. The place of the scripture which he read was this, He was led as a sheep to the slaughter; and like a lamb dumb before his shearer, so opened he not his mouth: In his humiliation his judgment was taken away: and who shall declare his generation? for his life is taken from the earth. And the eunuch answered

Philip, and said, I pray thee, of whom speaketh the prophet this? of himself, or of some other man? Then Philip opened his mouth, and began at the same scripture, and preached unto him Jesus. And as they went on their way, they came unto a certain water: and the eunuch said, See, here is water; what doth hinder me to be baptized? And Philip said, If thou believest with all thine heart, thou mayest. And he answered and said, I believe that Jesus Christ is the Son of God. And he commanded the chariot to stand still: and they went down both into the water, both Philip and the eunuch; and he baptized him. And when they were come up out of the water, the Spirit of the Lord caught away Philip, that the eunuch saw him no more: and he went on his way rejoicing."
Acts 8:26-39

So, what if there is more to this story? What if this detail was to give us another clue into where the Ark ended up, which in turn gives us more evidence of God's meeting with Jesus on Tana Kirkos Island? As history shows, there were a series of women warriors who held the office as independent rulers known as Kandakes, or Candaces, from around 345 BC to 314 AD in the area known as Ethiopia. These were spear holding and armored warriors who led the armies and if killed would leave the throne to her husband and son *(ref. 21)*.

In the time of Phillip, Queen Candice could possibly have been one of these warrior queens named Amantitere who ruled from 22 to 41 AD in Ethiopia. Regardless, this queen had appointed the Ethiopian Eunuch to be in charge of her treasury which gave him great authority in her kingdom. As Phillip arrived on the desert road known as Gaza, we see this Ethiopian sitting in his chariot confused about what he was reading in a large scroll from Isaiah describing the characteristics of Christ in chapter 53. Was the confusion of the Eunuch based on his visit to Jerusalem only to find it somewhat depressed? He probably had arrived in Jerusalem expecting to see some kind of celebration for the risen Savior. This may have led to his confusion when Phillip saw him pondering the words Isaiah.

Without a doubt this Ethiopian, along with his queen, would have read Isaiah 52 giving a look into the events leading up to, and possibly following Isaiah 53. But then a quick look at a verse in Isaiah 52 may give us a little more insight into the purpose of this Eunuch. Bob soon directed our thoughts to examine this point,

"Depart ye, depart ye, go ye out from thence, touch no unclean thing; go ye out of the midst of her; be ye clean, that bear the vessels of the LORD." Isaiah 52:11

Could this be what the Eunuch took as a sign that he was to go to Jerusalem to see if the treasures that Queen Candace was holding needed to be returned to the Christ at Zion? Scripture tells us that they are to depart and touch no unclean thing, which perfectly describes why Candice would have sent the Eunuch, simply because of his purity. And since this Eunuch was the keeper of the treasures of the queen, could the reference to the one who bears *"the vessels of the LORD"* be a direct tie to the Ethiopian's claim that they were holding, and possibly still are, the Ark of the Covenant and the Mercy Seat?

As we know, as noted in the Scripture, the Ethiopian accepts the salvation through Christ, and Phillip takes him down into the water to be baptized. Once realizing that the Mercy Seat was not needed as of that time and with his new found assurance, the Eunuch returned to Ethiopia rejoicing and telling others of his conversion. This again raises a question why the detail of this story, both in Acts 8 and Isaiah, if there is not more meaning to it. If it were only needed for a salvation and baptism experience, much of this Scripture would not have been needed. Maybe the Lord had more to share than we are given to believe. Maybe this gives us a closer glimpse into the importance of this ancient land of Ethiopia once again.

Hidden where no one would care to look

Looking back on this ancient land of the northern highlands of Ethiopia, we can see why this would be the perfect hiding place to one of the Bible's greatest treasures. The difference from this treasure and any other is that the Ark of the Covenant played such a great role in the Old Testament, and then is simply gone and mysteriously not mentioned again. That is, until the time that it will be needed again. So, why would Ethiopia be the perfect place to hide this mysterious artifact? The answer is simple, so no one would desire to pursue it in a land that is so diseased and where their waters are either dried up or spoiled and death is somewhat common place.

Ethiopia has continually experienced famine that seems to hasten the demise of these people through not only starvation, but also disease. Over seventy seven-percent of this nation is starving and below the poverty line. Malaria runs rampant in this country, and due to the lack of resources, people are unable to acquire the medication to treat or prevent this disease that is spread through the abundance of mosquitoes. Only seventeen percent of the nation has access to clean drinking water which is directly due to the lack of rain in this arid region and many others in Ethiopia. With that statistic added to the widespread cases of HIV, tuberculosis and other diseases, it doesn't take long to realize this is not your ideal site for a family vacation.

One of the most interesting statements that I heard came from a young man I spoke with in Axum, and then Bob reiterated those comments in his study and discussions; the Ethiopians in this region have accepted that this is the way it has to be in order to be the keepers, or custodians, of the Ark and Mercy Seat. They believe that they have been given this task and must suffer in order to be somewhat left alone until the appointed time has come. To the people of Axum, although there is some tourism created there from primarily outside sources, they would just as soon be left alone to worship and guard what they claim God has entrusted to them.

Chapter 13

PUTTING THE PUZZLE TOGETHER

Theories run wild in a situation like this. Without actually being there we may never know what actually took place during Jesus' time on the mysterious island of Tana Kirkos. But, without a doubt, we can assume some things by reviewing the facts of this miraculous journey. Just as Jesus said when the Pharisees asked Him to denounce His followers from crying out His Holiness in Luke 19:40, *"I tell you that, if these should hold their peace, the stones would immediately cry out,"* I can't help but feel that the incredible faith of the "keepers of the secrets" of the land of Ethiopia will once again cry out the holiness of Christ under the trust that God has bestowed on this land. Yes, they are swept in poverty, disease, and famine while laying claim to some of the most incredible artifacts and untold secrets that the world has ever known. I also believe that Ethiopia holds many more secrets that haven't even been imagined.

As we examine the pieces of this jigsaw puzzle, we discover many interesting, amazing and just plain and simple stunning facts crying out from the stones in the ancient land of Ethiopia. In order to comprehend the complete implications of this story, let's take some time to put these facts together so we can sense the magnitude of the evidence.

The Angel's Warning To Flee:

Once again, the story begins with the appearance of the angel to Joseph in Matthew 2:13 to warn him of the impending "slaughter of the innocence" with King Herod's fear, search and massacre of all the male children from the ages two and under. With fear and determination, Joseph follows the request of the angel and packs up his family and journeys to the preset trail for them into Egypt.

Discovering Egypt:

The incredible detailed history of the flight of the Holy Family into Egypt can be found in a series of recorded documents along the trail of this ancient land, as well as in the

Coptic oral traditions passed down through the ages. The trail is specific and carries a great deal of merit considering that many of the churches, ancient paintings and historic accounts date back centuries. Many traditionalists struggle with anything that doesn't have a firm foundation in the Scripture. I guess that's what changed my thinking along this trail. Sometimes God will work through those who may believe, but don't fully understand. That's exactly what He did with me. I was a traditionalist who, through a window of doubt, had a glimmer of hope that God would share a secret with the rest of the world through that hope. I still follow many of the old beliefs, but have realized that when God has something to reveal, He will change the minds of those He has chosen to complete the tasks. This all came true for me when I began to study the trail of the Holy Family in Egypt and then to the journey through the hidden lands of Ethiopia. I learned that we can't take what man always says as the gospel, but when God is involved, we have to keep our minds open to the heavenly possibility.

Venturing South:

In one of the most telling twists in this story, we see the Holy Family, after a montage of stops in northern Egypt;

make a turn to the south by meeting up with the Nile River and charting a due course into what is known as Upper Egypt, which is actually to the south. Even when questioned, historians, priests and monks in Egypt really have no explanation for this turn other than the pending threat of Herod's soldiers finding their location. Understandably, in the northern regions of Egypt that makes some sense. But when you look at the facts that Herod didn't even know who he was looking for, thus the slaughter of possibly thousands of innocent children (Matthew 2:16), there's a great deal of suspicion why the Holy Family necessarily headed due south and continued on the journey. This should raise an eyebrow and give you an "Ah ha" moment to this question.

Following the Great River:

Another interesting fact is that if Herod's soldiers would have been hot on the Holy Family's trail, wouldn't it make more sense to move away from the populated areas of the Nile River and disappear into a remote village off into the desert or an oasis region in Egypt? But, as we have learned, that's not what happened. They traveled from town to town with Jesus making known His entry as idols fell, and some of the "gods" worshipping priests became fearful at His appearance. In other words, there was no hiding that Jesus was who He claimed and Mary and Joseph didn't seem to hold it back, either. Even their arrival in Muharraq should raise red flags because of its far southern location in Egypt. Why would God allow them to travel, strain, and venture so far away from home in order to be called back when they arrived there? Could there have been further instructions? And why was it the Nile they stayed so closely to? The answers seem overwhelming, but very likely that the Holy Family may have stopped in Muharraq, liked it, and then returned to it after venturing down into Ethiopia for the prophetic meeting.

Ethiopia's Treasures:

The treasures that deal with the Ark of the Covenant, the Mercy Seat, and the so called missing pieces seem to fit together perfectly. And what is most interesting is that it appears as though the Ethiopians themselves haven't even put the pieces all together, nor do they feel they need to. They seem to take their divinely appointed commission for granted, in some ways. We find that they are guarding a lot more than the Ark, a missing link to some of the mysteries of Christ not shared in the Scriptures. Maybe these are the untold stories that John refers to in 21:25. Let's look at some of these treasures together:

Trumpets:

In Axum, Ethiopia they claim to have the ancient pounded silver trumpets of Moses mentioned in Numbers chapter 10. These trumpets were used to calling the people to the tabernacle when God had something to tell them. These trumpets were blown to announce the entry of the Ark to a new location and will one day sound again to announce its return to the third temple, (Isaiah 18:7). Today the Ethiopians claim to have these trumpets housed in the treasury in Axum. The second temple was prophesized to be destroyed, and it was in August of 70 AD by the Roman general, soon to be Caesar, named Titus. The trumpets that we actually saw and sketched out match perfectly to the trumpets described in the Bible. When the second temple was built the ceremonial items that had been taken were replicated and placed in their respective positions in the new temple. We can compare these trumpets to the replicas that Titus marched through the streets of Rome after his destruction of the second temple, as carved in the Arch of Titus that still stands in the Italian city today, we see a perfect match to the trumpets in Axum.

This event was horrific and vividly depicted by Jewish historian Flavius Josephus, who was in Jerusalem at the time of the capture of the city and when the demise of the second temple took place in flames: *"The Romans, though it was a terrible struggle to collect the timber, raised their platforms in twenty-one days, having as described before stripped the whole area in a circle round the town to a distance of ten miles. The countryside like the City was a pitiful sight; for where once there had been a lovely vista of woods and parks there was nothing but desert and stumps of trees. No one - not even a foreigner - who had seen the Old Judea and the glorious suburbs of the City, and now set eyes on her present desolation, could have helped sighing and groaning at so terrible a change; for every trace of beauty had been blotted out by war, and nobody who had known it in the past and came upon it suddenly would have recognized the place: when he was already there he would still have been looking for the City. (Ref. 3)"* Inside the temple were replicas of the original "treasures" which were taken and paraded by Titus in celebration of his victory. But, still today, the original trumpets are secured away in Axum, Ethiopia.

Ark of the Covenant:

Without a doubt the Ark of the Covenant is the most treasured and sought after artifact in history. The Ark of the Covenant with the Mercy Seat was built to house the Holy Law of God, the Ten Commandments, during the time of Moses. It rested in a series of makeshift tabernacles until it came to its home in the first temple of Solomon. After that the mystery of its whereabouts are shrouded in assumptions, theories, and turmoil. Biblical accounts disregard many of those theories, but in all, the clues leading the Ark and the Mercy Seat to Ethiopia seem to be indisputable. We would have to concur that the Ethiopians claim to be the guardians of this Holy object. This claim appears to be authentic, but only time will tell if their claim is divinely appointed.

The Guardian:

In Axum, and in Saint Mary of Zion Church, the Guardian of the Ark lives his life to worship and to guard this Holy object they claim to have. Do they have it? I can't answer that, and no else can either except for the Guardian himself. He is the only man alive that is perfect enough to be in the presence of the Ark and will guard it with his life until he chooses another Guardian at his passing. This man, considered to be pure and holy, surrenders his all inside the fence of this small church in order to pray and protect their claim. He seems to have a sadness about him, but a peacefulness that surrounds him as he wanders the perimeter of the grounds. This should raise an obvious question why someone would give his life to guard something that wasn't there. Maybe this should make us think strongly about what they claim to posses.

Their Faith:

The people of Ethiopia, pledging that they have the Holy Ark and that God has entrusted them throughout the ages with secrets, have an enduring and constant faith. These people don't have a great deal else in life to consume their time, but they put to shame the western world's practice of faith and worship. As I stood on a cliff and oversaw the people day after day, coming out of the mountains to worship sometimes for hours, I couldn't help but feel embarrassed about the limited time we give to God daily. In many cases their worship begins in the early morning hours, when the kebero drums begin to pound and the chanting worship and prayers begin. When you see people plant their faces in the sand praying for hours on end you begin to feel the desire for God's own heart that these people routinely seek. Still today, the Ethiopians carry an Old Testament faith that is an example to the world in their dedication to Him.

The Forbidden Island of Tana Kirkos:

This mysterious island is one of the most amazing places I have ever witnessed on this earth. Hidden some three and a half hours from the city of Barhir Dar in the massive expanse of the Lake Tana, this Ethiopian treasure appears to have been the site to a Holy meeting with profound significance. The island was not only the resting place for the Ark of the Covenant for nearly 800 years until King Ezana came to claim it for Axum, but from all indications is the exact place that Father God chose to reveal wisdom to a young Jesus during their meeting on this Holy piece of rock. The intense secrecy on this island never really hit home until I met the Ethiopians in the Library in Cincinnati. Their shock I had been on Tana Kirkos to photograph these pages brought a new perspective to the term "forbidden." It was as though this place was intentionally shielded from the rest of the world, and in its own right, it really was. The world was not welcomed here, and I thank Bob Cornuke for leading us to this apparent last place of righteousness on earth.

Guardian monks:

Just as in Old Testament times, the treasures of the Lord have been guarded both from mankind and from time. The holy men who guard this island have been carefully chosen at a very early age to carry out the centuries-old duties as guardians of the secrets of this island. Only a select few are permitted to visit this island and even fewer may pass through the jungle to the area considered to be "Holy ground." Once chosen, these monks will never leave this island and give their lives to raise their crops, worship and protect its age old secrets. Their abiding faith and their dedication to serving the Lord is noteworthy. They live in small huts not much bigger than a closet, they suffer through ailments and injuries without medication, they spend each day of their lives tending to crops, and they worship long hours in the temple. Their endurance is phenomenal and their heart for the Lord is unmistakable. They take their God-given assignment as guardians very seriously, but will hospitably share a cup of coffee or a piece of injera bread as a symbol of friendship and peace. These men are warriors of their faith who have taught me a great deal about to my own worship to the Lord.

Eagles:

The first thing I noticed when I came to the island of Tana Kirkos were the two white and black eagles that stood perched seemingly to guard the island from way above. Bob Cornuke told me that over the years during his visits to the island, these beautiful birds have always stood watch in the trees. Ironically the eagles stand guard above the site where most of the ancient Holy events took place. It's relative to the Scripture in Isaiah 40:31 that states, *"But they that wait upon the LORD shall renew their strength; they shall mount up with wings as eagles; they shall run, and not be weary; and they shall walk, and not faint."* In other words, the secrets that this island has guarded for centuries will be revealed, but we must wait for God's timing. When the appointed time arrives, God will put those secrets on the wings of eagles and give them to the world. It's somewhat symbolic, but a vivid reminder that God gives us our portion in his perfect time.

Tomb of the Priest:

Also on Tana Kirkos is the tomb of the high priest who brought the Ark of the Covenant to this island. Just a few yards past the flat stone surface that held the altar of the Ark of the Covenant are two large boulders forming a "V" shape;

it is covered and filled in with smaller stones and makes up the final resting place of the high priest. This rocky grave is a reminder to the accuracy of every detail that Ethiopia has as evidence to their claims. Today, one can barely see into this grave containing the decayed remains of this priest and possibly pottery shards that accompanied him during the burial. Again, interesting to note is that the place where the Ark was lifted to from the water some fifty feet below is just a few steps past the grave site. This adds further validity to the ever-growing evidence on Tana Kirkos to this case. Just think the remains could well be the high priest who carried out the Holy atonement ceremonial in the presence of God.

Holy Slab of Rock where the Ark Rested:

On the cliff where the Ark is claimed to have rested for 800 years is a flat stone surface measuring nearly to the perfect length of the Ark of the Covenant. After digging up old grass off the slab, I had discovered four carved out indentions in the stone measuring nearly the accurate distance of the resting feet of the Ark of the Covenant. And the ever mounting pile of evidence in this research was growing. In this same place Jesus allegedly knelt, prayed and listened to His Father as He was filled with the wisdom for His future ministry. If this claim is truth, and it could very well be, then this particular spot on this island conceals undeniable historical importance and scriptural reverence. This site could well be the exact location where God the Father and God the Son met.

Tabernacle Sockets:

Just a few feet away from the slab where the Ark rested is the first of the large carved out sockets in the rock that held the poles of the Tabernacle described in the book shown to us by the Abba. These pole sockets would have held the tabernacle in place and they stretch up the cliff side. Bob's discovery of these pole sockets are crucial to the accuracy of the claims and secrets of this island. With the whisking winds across the lake and up to the top of this island there had to be some way of anchoring the tabernacle that is claimed to have rested here in place. This is another confirmation that what they claim as truth is accurate. Although time has caused many of the sockets to plummet to the waters below, the vision of a once Holy tabernacle resting here is a geographically and architecturally phenomena to imagine.

Sacrificial Stone Bowl:

When Bob Cornuke first explained this structure to me, it was incomprehensible until I researched further. Just a few yards from the resting spot of the Ark is a small tin building that appears near collapse. Inside is a large stone basin with a carved-out bowl within the rock which the Abba claims to have been the bowl that the blood was poured into for the sacrifice, while the Ark was located on the cliff. It's interesting in the sense that actually it would have had no other purpose for being there other than for something ceremonial. Bob was confident of its purpose, and I concurred; the evidence indicates this is where the high priest would have dipped the blood sacrifice for the atonement ceremony with the Ark of the Covenant.

Solomon's Temple Treasures:

Also on the island of Tana Kirkos are the long lost treasures of Solomon's temple, not treasures of wealth, but the treasures of the tabernacle and the atonement ceremony, the true treasures of the temple. The Abba unlocked the large wooden door to the mud, rock and thatch building they call their treasury to an elaborate array of artifacts that, again, would serve no significance to anything else except for what they claim. We were privileged to behold those artifacts while in this small treasury room. Every one of these items confirms that the Ark could very well have been entrusted to these people and even more specifically the sacred meeting on the island.

Breastplate of the High Priest:

"And thou shalt take the garments, and put upon Aaron the coat, and the robe of the ephod, and the ephod, and the breastplate, and gird him with the curious girdle of the ephod," Exodus 29:5.

One of the most shocking items in the treasury was what they claimed to be the remains of the breastplate that the high priest would have worn during his entry into the holy of holies for the atonement service. The cage like presumably bronze harness made of flattened metal fit over the shoulders and would have been the undergarment frame to the cloth with the twelve stones of the twelve tribes of Israel attached to it. If this is truly what they claim then it is an astonishing find knowing it would have been worn in the presence of God's appearance on the earth. Holding this artifact in my hands literally gave me chills, and one of the men on our exploration team said that my face turned to a flush white when the Abba handed it to me. In reality, much of what was handed to me in this small room put me in tears. Just the overall thought of what was in my hands and what we were in the presence of was taking my emotions to an unprecedented level. All I know is that this harness of the breastplate has been in the presence of God Himself, if their claims are true.

Sacrificial/Cleansing Bowl (laver/gomer) and Stand:

Again, a presumably brass or bronze, heavily tarnished, artifact that fits the description in the Bible of the bowl for the blood sacrifice, or more likely the basin for the priest to wash himself in, would have been put in for ceremony before entering the Holy of Holies.

"And the LORD spake unto Moses, saying, Thou shalt also make a laver of brass, and his foot also of brass, to wash withal: and thou shalt put it between the tabernacle of the congregation and the altar, and thou shalt put water therein" Exodus 30:17-18

"*And Moses took the anointing oil, and anointed the tabernacle and all that was therein, and sanctified them. And he sprinkled thereof upon the altar seven times, and anointed the altar and all his vessels, both the laver and his foot, to sanctify them*" Leviticus 8:10-11

This bowl rests upon a perfectly fitted metal stand that has collapsed over time due to the age and weight of the bowl. This was another artifact that has no other purpose except to serve the purpose they claim.

Aaron's Almond Bud Meat Forks:

"*And it came to pass, that on the morrow Moses went into the tabernacle of witness; and, behold, the rod of Aaron for the house of Levi was budded, and brought forth buds, and bloomed blossoms, and yielded almonds*" Numbers 17:8

One of the most incredible pieces that links these items in the island treasures to Solomon's Temple and even before to Aaron and the Tabernacle is the almond topped meat forks for the burnt offering ceremonies. These long two pronged meat forks contain the design of an almond flower at the top.

This is a direct witness to the theme of the almond bud that God blessed Aaron's staff with and later became a signature of Aaron's position. This is one of those pieces that you question all the details: metal giant fork on a remote island with an almond flower bud engraved into the top of the joint between the prongs. The appearance of the almond was a deciding factor confirming the authenticity of all the other artifacts shown to us on the island and directly dates these objects to the time of the original tabernacle.

Burnt Offering Hooks:

"And the priest shall put some of the blood upon the horns of the altar of sweet incense before the LORD, which is in the tabernacle of the congregation: and shall pour all the blood of the bullock at the bottom of the altar of the burnt offering, which is at the door of the tabernacle of the congregation" Leviticus 4:7

The meat hooks for the burnt offerings are again made of sturdy metal and have a collection of hooks on them in order to hang the meat that was prepared for the burnt offerings to the Lord. These are perfect examples of the detail that the monks wouldn't need to confirm their finds, but because of their overt innocence, add to the pieces of the puzzle. Many of the items aren't really needed to justify their claim but do lend themselves to the overall authenticity of the artifacts.

Incense Burners and Shofars:

*"And Aaron shall burn thereon sweet **incense** every morning: when he dresseth the lamps, he shall burn **incense** upon it"* Exodus 30:7

Many other items located in the treasury on Tana Kirkos Island aren't as equally spectacular, but they still support to the theory and claim of the existence of the Ark on this island. The incense lamps that hang unused for centuries are contained in this treasury, and since the Ethiopian monks burn incense during their worship, the smell permeating the robe of the Abba prompted a supernatural realization as we stood in their presence.

"And seven priests shall bear before the ark seven trumpets of rams' horns: and the seventh day ye shall compass the city seven times, and the priests shall blow with the trum-

pets. *And it shall come to pass, that when they make a long blast with the ram's horn, and when ye hear the sound of the trumpet, all the people shall shout with a great shout; and the wall of the city shall fall down flat, and the people shall ascend up every man straight before him. And Joshua the son of Nun called the priests, and said unto them, Take up the ark of the covenant, and let seven priests bear seven trumpets of rams' horns before the ark of the LORD...And it came to pass, when Joshua had spoken unto the people, that the seven priests bearing the seven trumpets of rams' horns passed on before the LORD, and blew with the trumpets: and the ark of the covenant of the LORD followed them... And seven priests bearing seven trumpets of rams' horns before the ark of the LORD went on continually, and blew with the trumpets: and the armed men went before them; but the rereward came after the ark of the LORD, the priests going on, and blowing with the trumpets."* Joshua 6:4-6, 8, 13

Another item resting on the shelves and decaying over the centuries due to damp musty air within the building, are the shofars horns. These long rams' horns were hollowed out and were usually blown for celebrations or for warnings of a pending approach of the enemy.

Place of Mary's Prayers:

High on the top of the cliff overlooking the treasury, the resting place of the Ark of the Covenant where her Son would have been praying with the Father, is the prayer rock of the mother, Mary. The interesting thought to this particular site is that you can see nearly the entire island, and islands far away from this post. This is also the location of the roosting black and white eagles standing watch over the "Holy ground." At the spot they claim that Mary watched over her Son as He prayed with God, they have a very small three foot by three foot tin building marking it as a holy site for their worship. The view from this point is spectacular, but the thought that Mary would have knelt here day after day as she prayed for Jesus is truly inspirational. If only the stone would talk to us now and reveal the events that took place here some two thousand years ago.

Where Jesus Knelt:

When the Abba said this to Misgana, and he translated for me, I was both in confusion and doubt. Documented with evidence from the Abba and based on the historical account in the Ethiopian Bible. I will never forget the numbness that overtook my arms and fingers as I fell to my knees, not knowing what had taken place on this exact spot: stone slab that the Ark of the Covenant had rested on for eight hundred years and the exact spot in front where a young Jesus had taken His position, just like a school boy taking his place at his desk, to learn from God's teaching and absorb the tremendous wisdom we read about in Luke 2:40. I couldn't help but ponder Jesus' reaction to the His Father when it was revealed that He would one day be the sacrifice for all mankind. It reminds me of the plea of Christ in the Garden of Gethsemane prior to the arrival of the Roman soldiers with Judas in Matthew:

"And he went a little further, and fell on his face, and prayed, saying, O my Father, if it be possible, let this cup pass from me: nevertheless not as I will, but as thou wilt" Matthew 26:39

One of the most inspiring moments in my life took place on this rock where some two thousand years earlier Jesus Christ spent three months and ten days in the presence of the Father Himself in this same location and on this same spot on this earth.

The Proof is in the Pages:

None of this would have been possible if it weren't having been for the kindness of the Abba and the timely approval of God to reveal the contents of the historical Ethiopian Bible in the treasury. Resting among the dusty old books, Bibles, and scrolls, lay a large ancient Bible containing a long guarded secret of the land of Ethiopia. As the Abba in Cincinnati referred to the contents as "a secret in Ethiopia for thousands of years and few people know about this, about Jesus and Maria coming to this place." By this time I was well aware that we had been given a great honor to be shown such a document. A piece of ancient history that was to be hidden away until someone was to either stumble across it, or it was to be shown to someone to tell the story. Somehow, that someone just happened to me. Like every person chosen by God for certain purpose, I, too, question "why me?" The journey, the events, the people are all pieces of a heavenly puzzle I find myself putting together.

John's Telling in John:

During the translation of the Scriptures one of the more shocking pieces of the text came when I learned that Jesus' disciple, John, was the author of the story and that John had requested this secret be taken to Tana Kirkos for safe keeping. The most telling revelation came when the Abba in Cincinnati clarified the eight and an half lines that warn about removing, changing or adding to the pages of the book. When I told the Ethiopian translators that these same warnings were in our Bible, they were quick to tell me that this was written before John wrote the ending to Revelation. Finally, the Abba added that the writings were written down by Prochorus, the deacon chosen in Acts 6, Peter's apprentice, and John's student. Prochorus is also the main ingredient to the connection of the warning both in this book and in Revelation since he was the one who most likely assisted John on the island of Patmos when the final book of the Bible was recorded.

"For I testify unto every man that heareth the words of the prophecy of this book, If any man shall add unto these things, God shall add unto him the plagues that are written in this book: And if any man shall take away from the words of the book of this prophecy, God shall take away his part out of the book of life, and out of the holy city, and from the things which are written in this book. He which testifieth these

things saith, Surely I come quickly. Amen. Even so, come, Lord Jesus." Revelation 22:18-20

Knowing that this story was given to John by Mary some time after Jesus' crucifixion, resurrection and ascension added even more credibility to the overwhelming evidence mounting in this story.

Clearing the Path:

Evidence continued to mount as each step along the path was cleared leading us on an unprecedented adventure. The entire trip was ordained to make sure that it presented to the world properly. I see God's hand in each and every piece of this complicated, but revealing puzzle. This journey started with a family outing in Gatlinburg, Tennessee, to an unusual phone meeting to order a DVD, to an eventual return call back and then finally to an invitation to be a part of an exploration that would take us by faith to an ancient world. That was followed by fundraisers, meetings with people in need, revelations on the cliffs of Axum and then eventually arriving in Barhir Dar and traveling by boat to Tana Kirkos Island. But the clearing didn't stop there. Each step was a "dusting off" of the walkway that we were being led to take.

There wasn't a hedge along the path except when I would take my eyes from it.

Confirming the Translation:

Another part of the "clearing" was the way the translation of the documents came together. After returning, one of our biggest prayers was to be able to find someone to help us translate the ancient Ge'ez language so we could grasp the meaning of the Scriptures. Since Ge'ez is a language like none other, we needed an authority who could accurately translate the contents. Surely as a result of divine intervention, a revered Ethiopian Abba met with us and obligingly transcribed the words of the pages and removed the blinders from our eyes and confirmed the claims of the Abba on Tana Kirkos Island. This marked the beginning of an enduring friendship that seems to complete my round-the-world journey.

The Revelation to the World:

With verification of all our discoveries and revelations, it was quite evident to me that we needed to share this story with the world. Even though this has been a well-guarded hidden secret to the whereabouts of Christ as a boy, it is also testimony to the fact that when God wants something revealed, He will clear all obstacles to make it happen. This information clarifies multiple questions about the travels of the Holy family. I guess what makes this a more incredible discovery is that many so called historical texts have been written and dismissed because of man's denial that any truth beyond the Bible would be unfounded. Just as the translators of the King James Bible only included the texts that were consistent with the overall Word of God, the same should be discerned with the many newly discovered hidden historical manuscripts to their confirmation, not replacement, of the Holy Scripture. This takes us back to John 21:25 telling us that there was much more that had yet to be recorded, or that had been recorded but not revealed. It did not say that they weren't written down at all. It's the same with the Disciples of Christ and their historical demise. We have historical record of the fate, many times a tragic one, of the Disciples of Christ. Their deaths were recorded and the historians accept their story in historical documents, but they weren't included in the Holy Scriptures. Therefore, why can't we accept that there was more to Jesus' life recorded than what the Bible tells us? The Scripture gives us what we need for our lives, but historical documents give us just that, the history. On the top of a mysterious island in the middle of a massive lake in the Ethiopia, a place mentioned many times in the Scripture, carried an unmistakable secret in history that needed to be revealed, and now it has been.

The Scriptures Lead Us

And finally, within the Scripture, we see a young boy who was born to the world in order to save it from the sin that has overcome it, and then disappears from the pages until we find that He has been filled with wisdom, knowledge and the grace of God. Then again, Jesus is gone from the Scriptures. Now just maybe with this miraculous find, we have a better idea what happened following His trek into Egypt and beyond and how He obtained the knowledge He would need for all future events.

The Last Piece of the Puzzle in Place

In many ways we need to open our eyes to what God is revealing all around the world. In the middle east there have been more locations and artifacts found in the last several years than any other time in history. So quickly and plentiful, in many cases, that we don't hear about them all because the news would be filled with stories of new finds all the time.

Besides, that's what the enemy wants, to veer you away from newfound historical biblical truth to keep you ignorant of God's fulfilling prophecy. What better way to cause question to those who don't believe in Christ in the first place! If Christians don't believe, then why should anyone else? Maybe that's what we need to really consider in this world of confusion today.

Chapter 14

WERE THEY REALLY LOST YEARS?

Were the years of Jesus as a boy really lost? Did He really do anything of significance during those unmentioned years of the Bible? Were His only important and God-like episodes written in the Holy Scriptures we read about today? I believe along with many others I've encountered since I began this research, there is no way that Jesus' fullness of knowledge in Luke was a "snap your fingers and you're wise" occurrence.

"And when he was twelve years old, they went up to Jerusalem after the custom of the feast. And when they had fulfilled the days, as they returned, the child Jesus tarried behind in Jerusalem; and Joseph and his mother knew not *of it. But they, supposing him to have been in the company, went a day's journey; and they sought him among their kinsfolk and acquaintance. And when they found him not, they turned back again to Jerusalem, seeking him. And it came to pass, that after three days they found him in the temple, sitting in the midst of the doctors, both hearing them, and asking them questions. And all that heard him were astonished at his understanding and answers. And when they saw him, they were amazed:"* Luke 2:42-48a

Referring back to John's comment in chapter 21 of his gospel, that there was so much more that Jesus did that was not included in the Scriptures we have to assume that there was more to the story waiting to be told. John is specific in saying, *"there are also many other things which Jesus did,"* that gives us the direct knowledge that much more happened in Jesus' life that was not mentioned. Is it because we couldn't comprehend or He didn't want us to know until now? I'm still not implying that the Bible is incomplete. Rather it is exactly what God breathed into men to be written and presented for our understanding and knowledge for the perfecting and education of those who believe in Him. I believe that, just as with any other figure and event in history, there is much more to the story than is written. And with the obvious "omission" of the Scriptures nearly thirty years of Jesus' life, we must conclude that He didn't just disappear and then return for three years to minister in the Holy land. We have to believe that many other events took place that weren't included in the gospels, but in time would be revealed for the world to know.

Webster's defines the word "lost" as follows: "taken away or beyond reach or attainment; no longer visible; hopelessly unattainable." With that definition as our starting point, we could speculate that God possibly with held these years from our knowledge and from beyond our reach. But the word "lost" can also be applied in a variety of ways. For example, "I was lost, but now I'm found," applies here. The knowledge of many biblical sites was once lost but now found for us to view first hand and to pay homage at as Christians. But what about items that may never be seen by the world, such as these lost documents from Ethiopia? Yes, I have had the opportunity to view historically valuable hidden pages written by John that disclose the time lapse in Jesus' life. We were allowed to photograph those pages, but the world may never actually see these documents on display as you may see the Dead Sea Scrolls exhibition or the

Declaration of Independence in Washington D.C.'s National Archives Museum.

Re-examining the Scriptures, this same thought comes to mind regarding the Ark of the Covenant and its whereabouts. The theories of its location are in most cases just that, a theory. From a biblical perspective we see justification for the Ark's current resting place by looking ahead to the future. At some time in history the Ark, or Mercy Seat, will re-emerge to fulfill its purpose with Christ. Looking at the reference as "hopelessly unattainable," we are sure to see that this is simply not the case when it comes to God, the Bible, and anything else to do with it. As we read in Mark 10:27 in Jesus' own words, *"With men it is impossible, but not with God: for with God all things are possible,"* we simply find that if God is involved, we can't assume that these documents, years, and artifacts are lost. With God, nothing is ever really lost.

For nearly two thousand years, the trek of the Holy family was thought to be a myth or fable. Today, many still think they are. Even Paul Perry, author of the best-selling book *Jesus In Egypt*' remained under strong conviction whether the information he was uncovering about this trail through Egypt was factual. It was not until a vision, or a flash of light appearing from above a chapel in Egypt captured in a photograph taken by Perry, that opened his eyes to confirm his research and beliefs. Through nearly two thousand years of hidden secrets about the Holy family's trip through Egypt, the documents we were shown weren't lost after all. They had merely been sealed from the rest of the world, documents that would confirm and elaborate on the Holy family's journey. The same theory applies to the life of Christ. Somewhere the story of His untold days on this earth are safeguarded until the time is right to share them with the world.

Jesus' long and dangerous journey to Ethiopia explains why the Holy family continued to travel the Nile River southward to the bottom of Egypt with eventually ending in Lake Tana and more precisely, Tana Kirkos Island. This stressful but divinely appointed journey precluded the events that would shape Jesus' ministry. Their presence on Tana Kirkos Island two thousand years ago paved the way for the historical discoveries today, including - the time period Jesus was on the island, the accuracy of the tabernacle, the persons included in the journey and their locations, and that God's preparation of Jesus for the years to come. It was a "crash course" for Jesus in what was to become the greatest ministry in the history of mankind.

When we began our research, we had no idea that such important documents and writings would be revealed to us. We were in search for clues to the whereabouts of the Ark of the Covenant, as well as to reach the people of this country. In turn, God trusted us to uncover the once lost story of a young boy who made a long strenuous journey with His family to arrive at a mysterious island for a biblical meeting beyond comprehension. As the Abba said, this has been a secret of Ethiopia for nearly 2,000 years. As with many artifacts and sites that are being uncovered in and near the Holy land, the secret of Jesus' meeting with His Father was no longer to be a secret. As it is with everything in God's revelation, it was time for the Heavenly Summit to be revealed to the world. It was time to begin filling in even more of the mystery of Jesus as a boy.

The Reference to Ethiopia in the Bible

Even though the newly discovered information about Ethiopia is overwhelming, biblical and historical texts support the findings. From the Garden of Eden, to Moses' wife, to Jesus in Ethiopia, to the Eunuch of Ethiopia, the evidence and clues that the Bible provides, supported by archeological finds and historical writings, give us overwhelming evidence that Ethiopia plays more importance than what we have been taught over the centuries. However, to defend those who have been teaching, tradition has always focused on the area in and around Jerusalem and not beyond its borders.

Could Ethiopia have been host to a heavenly summit? Is this the place that God has entrusted to guard the Ark of the Covenant? There has to be much more to this story than we may have first realized, which opens up opportunity for future explorations and revelations.

The Biblical Record

Using the Bible timeline in Luke 2:7, we note the birth of Jesus in Bethlehem, *"And she brought forth her firstborn son, and wrapped him in swaddling clothes, and laid him in a manger,"* which explains the basis of the journey by Mary and Joseph to the city of Joseph's lineage. The census, as ordered by Caesar Augustus, was probably God's way of getting them to make the trek in first place. Just as He does in our lives when He needs something accomplished, the Lord will use other means to get us there and then reveal the true need after we have arrived. Following the visit by the shepherds, Jesus was circumcised (Luke 2:21), a few days later Mary and Joseph took Him to Jerusalem to present Him to the Lord, a custom at that time, and is still a custom in some beliefs today. This young couple stood in wonder as Simeon prophesized their Son's future.

Following this event and the timeline Matthew two tells about wise men, who traveled from the east into Jerusalem. King Herod learned of their journey and their purpose to follow the star that was leading them to worship the King of the Jews;

"Then Herod, when he had privily called the wise men, inquired of them diligently what time the star appeared. And he sent them to Bethlehem, and said, Go and search diligently for the young child; and when ye have found him, bring me word again, that I may come and worship him also. When they had heard the king, they departed; and, lo, the star, which they saw in the east, went before them, till it came and stood over where the young child was. When they saw the star, they rejoiced with exceeding great joy." **Matthew 2:7-10**

As we know, there was more to Herod's bidding than to worship Jesus. He played the wise men as pawns in order to find the location of Jesus, who was a threat to his reign as king, but God redirected them not to return to Jerusalem because of Herod's ulterior motive (Matthew 2:12). We also learn in this portion of Matthew that when the wise men arrived in the house, an obvious move from a stable, they worshipped a young boy and not the small baby as seen in Luke chapter two:

"And when they were come into the house, they saw the young child with Mary his mother, and fell down, and worshipped him: and when they had opened their treasures, they presented unto him gifts; gold, and frankincense, and myrrh" Matthew 2:11

From this point, with Herod's pending decree to have all the young children two and under killed, we must assume that Jesus was around the age of two at this time. It was here that Joseph receives a Holy visit by an angel:

"And when they were departed, behold, the angel of the Lord appeareth to Joseph in a dream, saying, Arise, and take the young child and his mother, and flee into Egypt, and be thou there until I bring thee word: for Herod will seek the young

child to destroy him. When he arose, he took the young child and his mother by night, and departed into Egypt" Matthew 2:13, 14

This is where events become vague and speculative. From around the age of two, presumably, until the approximate age of twelve we are lost in the world of speculation to the whereabouts and activities of Jesus and His family. Or, at least, that has been the thought for nearly 1,900 years. Just as we must look into historical records and recorded eyewitness accounts of some of the most popular figures in history, we must do the same for the miracle- filled life of Jesus Christ. We know they went into Egypt, so Egypt was the perfect place to look for clues about their trek throughout that country. Although some of the recorded events may appear exaggerated, we have to remember that we are talking about Jesus, the Messiah, and that anything is possible through God.

We now return our attention to the oral and written historic text in the Egyptian accounts and the Coptic biblical historic record to make sense of the southward journey of the Holy Family through Egypt and then farther down the Nile River to Ethiopia in Lake Tana and finally to Tana Kirkos

Island, where Jesus met with God, Himself. With the documented writings hidden on the island of Tana Kirkos, recited by Mary and recorded by the Apostle John, then delivered by Prochorus to the Ethiopians for secret storage, we can see God's accuracy and timetable beginning to take shape. After Jesus grew in wisdom and understanding, the Holy Family made their way back to Egypt and took up shelter in Al Murharraq awaiting the call to return home. Scripture tells us that an angel came to Joseph in much the same way as when he was told to go into Egypt:

"But when Herod was dead, behold, an angel of the Lord appeareth in a dream to Joseph in Egypt, Saying, Arise, and take the young child and his mother, and go into the land of Israel: for they are dead which sought the young child's life. And he arose, and took the young child and his mother, and came into the land of Israel. But when he heard that Archelaus did reign in Judaea in the room of his father Herod, he was afraid to go thither: notwithstanding, being warned of God in a dream, he turned aside into the parts of Galilee: And he came and dwelt in a city called Nazareth: that it might be fulfilled which was spoken by the prophets, He shall be called a Nazarene." Matthew 2:19-23

In these five verses we see Joseph's encounter with the angel once again giving him the all clear to return to Israel. But he was apprehensive about Jerusalem, however, because Archelaus, Herod's son, was now reigning king. So Joseph packed up his family and traveled back into Israel, bypassing Jerusalem, and traveling into Galilee back to their new home in Nazareth. This agrees with Luke 2:39, *"And when they had performed all things according to the law of the Lord, they returned into Galilee, to their own city Nazareth."*

This recorded return to Nazareth confirms my own thoughts, specifically noted while sitting on a rock outside the ancient treasury on Tana Kirkos Island.

"And the child grew, and waxed strong in spirit, filled with wisdom: and the grace of God was upon him" Luke 2:40

The Holy Family had to take this detour for Jesus to receive His strong spirit and wisdom. After their return to Israel, the young Jesus astonished some of Jerusalem's most educated men with His Spirit-filled conversation.

"And it came to pass, that after three days they found him in the temple, sitting in the midst of the doctors, both hearing

them, and asking them questions. And all that heard him were astonished at his understanding and answers" Luke 2:46, 47

At the age of twelve we know He confounded those who were in His presence. Some eighteen years later He did the same when He appeared to the world as the Son of God, the Savior, the Lord of Lords and the King of Kings. Wouldn't we like to know about those interim years not mentioned in the gospels? I believe it's only a matter of time before God reveals those to us. Even the Ethiopians who guard this sacred evidence do not understand the extent of their Holy treasury.

So...What's Next?

When we really step back and examine the whole picture of what will be the next piece of the biblical puzzle to be unearthed, we must first allow our hearts and minds to be opened to anything. In the last ten years, we have seen a mounting list of historical finds that line up with events in the Bible: the anchors from the Apostle Paul's shipwreck; the etchings confirming the existence of the House of David; various writings and scrolls uncovering of the Siloam Pool; the continued research on the locations of

Sodom and Gomorrah; the finding of the ancient boat in the Sea of Galilee; confirming evidence to the true site of Mount Sinai; the church of Mary Magdalene; the location of Herod's tomb, and so many others too numerous to mention. The discoveries are coming so quick that publications can't keep up with them.

To answer the question, "What's next?" I truly believe that the doors have been unlocked to what God will allow to be found. Just as we stumbled upon these historic and revealing pages of Jesus' early life, the same is true with the future unearthing of what only God knows will be next. What I think is more important is that we keep our minds open to the new discoveries and let our discernment help us to find the truth along the way. I prayed diligently about the authenticity of this information. I spent hours upon hours in solitude and prayer deliberating over the findings and the documents that we were shown. After putting all of the evidence on paper God then opened up the path and threw bread crumbs along the trail for me to either deny or to pick up and move forward.

This has truly been an interesting journey, one of unsuspecting revelation and struggle. No doubt there will be some who question our evidence and rightfully so. I was, and still am in many cases, one of those who doubt. But as I have heard many say before, until you have been there, felt the hand of God on the journey, seen doors opened to clear the path, and felt the joy that comes when following His lead, then I encourage you to at least consider the worthiness of the evidence. Just as I had doubted the opportunity of salvation through Jesus Christ, I am truly joyful that I opened up my eyes and my heart to His eternal assurance. My only regret is that I didn't do that earlier in my life. I pray this analogy enlightens your understanding.

In the Holy Bible in the book of Luke, Jesus sat with His disciples, and many others who gathered with them, and He began to share a parable about the sower of the seed. This popular parable pinpoints where our faith lies and is the foundation of the truth that our roots are secured in. When Jesus came to the end of the parable, He gave the answer to understanding the mysteries that the Bible holds, and foretold the mysteries that are still to come:

"And he said, Unto you it is given to know the mysteries of the kingdom of God: but to others in parables; that seeing they might not see, and hearing they might not understand.

Now the parable is this: The seed is the word of God." Luke 8:10, 11

In this passage Jesus tells us that those who truly believe in Him and fill themselves with the knowledge and Word of God and His teachings, will have a better understanding in His revelation of the mysteries of the Lord. But those who don't believe in Him will have a great deal of trouble understanding and believing. God's Word lays the foundation and trusting in Him will open up the doors to a greater and fulfilled understanding. I sometimes wonder what I missed during my years of struggling with the truths and not putting my faith in God's Word. Once again using the comments of Bob Cornuke, "People say show me and I'll believe, but God says you believe and I'll show you the truth."

In many cases we find that these holy sites and newly discovered biblical un-coverings have long been revered by those living in their proximity, and they have been paying homage to these sites, or secrets, for centuries. Just as the truth of Jesus' arrival in Ethiopia has been revealed in this book, so has this secret been guarded by the Ethiopians for thousands of years until now. God had chosen their land to house His secrets, only to be revealed in His timeline. Were the years of Jesus as a boy, and those after His encounter in the temple teaching really lost? I would have to say no, they aren't really lost, but they are simply hidden away until the appointed time where God feels that we are ready for the next revelation. So, what's next? That is a question that only the Lord can answer, but when it is revealed to us, one thing is for sure; it will be spectacular and aligned with every truth He has ever given us.

Chapter 15

OUR ADVENTURE BEGINS... AGAIN!

It all started with a phone call and ended up with a revelation of a lost story about what truly happened to Jesus during His time as a boy. We know the Bible only mentions that the Holy family was told to go to Egypt, but, through a tremendous amount of research and a miraculous experience, we now know that the journey of this young boy, who would one day save the world, was much more than a retreat. During this adventure, I discovered my skepticism, a critic to anything that was out of the traditional box. Then, I looked at my own life and realized that I myself was far from the tradition. I was raised a Catholic, saved a Pentecostal, and baptized a Baptist. Yet, I was over my head in the tradition of man in the church. I was told "if we don't teach it and the Bible don't state it, then it ain't true." Then Abraham Lincoln wasn't our sixteenth president, I guess. Now I have realized how ridiculous that really is. With a statement like that we start to realize how so many people today can't get a grasp on the Christian faith. We start to realize that statements like that bring no credit to God or the teachings of Jesus. Anyone blind enough to overlook the fact that we have nearly thirty years of the life of Christ untold in the Bible, and believe that absolutely nothing happened worth telling during those years, has just compromised the power of our Lord and Savior.

I don't believe in compromising the Word of God in any way, shape or form. At the same time I don't believe that we must dismiss the obvious when God is calling us into another direction or journey in our lives. Many believers miss the opportunities and blessings that God bestows on them because their short minded, traditional theories and beliefs hamper them from seeing the truth before them. When God opens the gates and sweeps the walkway for us to run the race, as the Apostle Paul referred to in Hebrews 12:1b-2a, *"let us lay aside every weight, and the sin which doth so easily beset us, and let us run with patience the race that is set before us, Looking unto Jesus the author and finisher of our faith*," then we are to follow that path until the mission of the Lord is completed. Sometimes we are not privileged to know the outcome in the beginning, but when our eyes are opened, we will see the glory of His will for the adventure in the end. And then, there are times, where we may never know why He sends us on a pathway in our lives.

This adventure to a forbidden island, to a people scattered and peeled, to a land of beauty and a land of pain, was set before us and God Himself swept the path clean for us to complete the journey. Now that we have completed this portion of the mission, you begin to see a clearer picture that there is much more to come. Much more to be revealed in time and it will all come in "*such a time as this*."

The Nile Route Revisited

Just as the Bible reminded me of the fingerprints we have on so many we come in contact and witnessed with, I couldn't help but think that the fingerprints of Christ are on us all. The story told in this book becomes clearer each and every time I look over this information. Looking back at the outset of this Holy journey to the south makes so much more sense to me now that I have the facts and path for reasoning.

I guess I never questioned the trek of the Holy family into Egypt because that's what tradition has always said. "They went there" and "They stopped here" and then "They came home from there." Unfortunately, that assumption has cheated us out of a great deal more of the historical accounts of Jesus for centuries. Joseph received a call from above to get this family out of Israel into the safety of Egypt. It has become evident that the call was to remove them from their surroundings and threat of Herod to make their vision clearer to the need at hand. Consider this; the turmoil with the virgin pregnancy, ridicule, and now the possibility of death due to soldier's hands, would have overwhelmed the human nature of Joseph and Mary. Even if Jesus Himself would have told them in the Holy Land that they had to go to Egypt to begin a tiresome journey to the south because He had to meet with His Father, I don't think that they would have been able to receive the thought because of the confusion surrounding them. Thus, the Holy Family had to be removed from their chaotic circumstances in order for God to alert them to the next leg of their journey. This was a journey that, until now, had been kept in the confidential file, or more accurately in an old musty rock and thatched building.

The trail begins in the northern Egypt and swirled through a series of unwelcomed stops throughout the area. Then, for no apparent reason, the Holy Family turns to the Nile River and heads south. Not just a few miles south, but hundreds of miles to the southern town of Muharraq. As mentioned before, it was believed for decades that the family of Jesus tarried in this place until Joseph received the calling of the angel to return to their homeland of Israel. After reviewing this, and pondering on it, you have to ask yourself the question "why that far to the south?" Or was there more? Just because the Bible mentions only Egypt doesn't mean that this is the only place they traveled to. As a matter of fact, it was probably only mentioned because the rest was not needed, as stated in previous chapters, to the path of righteousness for the believers. Regardless, it is still important historical information that, until now, has been hidden away until God felt the time was needed to reveal it.

The Holy Family's trail didn't stop in Egypt, which explains the long trek to the farthest regions of the country on the Nile River in the south. We now know that it was all preset to the place that was pure and undefiled located at the very end, or beginning, of the great Nile River, to an island that God had hidden away from the sin of the world where His Holy Ark was taken for His appearance on the throne to meet with His Son presented in human flesh. Tana Kirkos Island was the setting for the Holy education of Jesus as a child in order to give Him the knowledge for the future events to come. When the evidence is laid out, we begin to see how this all comes to the pinnacle of a perfect story. After Joseph received the approval from the angel to return to Israel, the very next happening we see upon their return is that Jesus had been highly educated and is teaching in the temple. Mary and Joseph understood that He was doing what He had been told to do.

"And when they found him not, they turned back again to Jerusalem, seeking him. And it came to pass, that after three days they found him in the temple, sitting in the midst of the doctors, both hearing them, and asking them questions. And all that heard him were astonished at his understanding and answers. And when they saw him, they were amazed: and his mother said unto him, Son, why hast thou thus dealt with us? behold, thy father and I have sought thee sorrowing. And he said unto them, How is it that ye sought me? wist ye not that I must be about my Father's business? And they understood not the saying which he spake unto them." Luke 2:45-50

Jesus: In Ethiopia

After His education with His Father on Tana Kirkos Island, the Holy Family more than likely returned to Muharraq, in Egypt, to await God's confirmation that it was time to return home. From there another story is hidden to be revealed later. Now that we have a clearer focus on the early years of Jesus' walk on the earth and where His purpose was given to Him, the question comes to me, "Where is the history of the teenage years and into His twenties?" Even though it has been called the "Lost Gospels," we now understand that this may not necessarily be the truth. This could be another one of man's traditions to keep us from locating or searching for the missing years. Just as God laid out this path perfectly to find the apparent missing link to Jesus' childhood and education, it is also likely that the remaining years of Jesus' ministry on this earth during those so called "lost years" are awaiting someone to stumble upon in that time when God sees necessary.

Reflecting Back On Timkat

During the two days that we witnessed the ceremony of Timkat in Axum, Ethiopia, it was somewhat a period of disbelief for me. We had seen so much, and God had revealed some incredible truths to me that I was literally unable to comprehend all that was going on at the moment. Reflecting back, my mind clearly looks at the Timkat in a double fold. First, to the time of King David during his attempt to bring the Ark of the Covenant to Jerusalem in a festive royal procession that led this Holy artifact into the city;

"So David gathered all Israel together, from Shihor of Egypt even unto the entering of Hamath, to bring the ark of God from Kirjath-jearim. And David went up, and all Israel, to Baalah, that is, to Kirjath-jearim, which belonged to Judah, to bring up thence the ark of God the LORD, that dwelleth between the cherubims, whose name is called on it. And they carried the ark of God in a new cart out of the house of Abinadab: and Uzza and Ahio drave the cart. And David and all Israel played before God with all their might, and with singing, and with harps, and with psalteries, and with timbrels, and with cymbals, and with trumpets." 1 Chronicles 13:5-8

The problem was that David was uneducated about the transport of the Ark when it was placed on the cart which caused the eventual death of Uzza by touching it to stabilize it, and the depression put upon the City of David. After the

Ark was taken to the house of Obed-edom, and they received many blessings from its time there, David realized his mistake. The Ark must only be transported by the Levites who carry it upon their shoulders. So, again we see the correct procession of the Ark of God being paraded through the streets to its resting place in the Tabernacle in Jerusalem;

"And the children of the Levites bare the ark of God upon their shoulders with the staves thereon, as Moses commanded according to the word of the LORD. And David spake to the chief of the Levites to appoint their brethren to be the singers with instruments of musick, psalteries and harps and cymbals, sounding, by lifting up the voice with joy...And Shebaniah, and Jehoshaphat, and Nethaneel, and Amasai, and Zechariah, and Benaiah, and Eliezer, the priests, did blow with the trumpets before the ark of God: and Obed-edom and Jehiah were doorkeepers for the ark. So David, and the elders of Israel, and the captains over thousands, went to bring up the ark of the covenant of the LORD out of the house of Obed-edom with joy. And it came to pass, when God helped the Levites that bare the ark of the covenant of the LORD, that they offered seven bullocks and seven rams. And David was clothed with a robe of fine linen, and all the

Levites that bare the ark, and the singers, and Chenaniah the master of the song with the singers: David also had upon him an ephod of linen. Thus all Israel brought up the ark of the covenant of the LORD with shouting, and with sound of the cornet, and with trumpets, and with cymbals, making a noise with psalteries and harps." I Chronicles 15:15, 16, 24-28

This same feeling was truly what the ceremony of Timkat felt like. Even though the ceremony represents (again according to whom you ask) the baptism of Christ in the River Jordan, and the overall pageantry resembles the procession of the Ark into Jerusalem. The trumpets are blaring, the harps and sistrums are clinging in a chant-like manner, and the large kebero bass drums keep the movement of the procession in motion through the streets.

When the monks exit the Holy Saint Mary of Zion church, they are clothed in full Levite ceremonial dress and the honored priest carries the replica of the Ark, or the Tabot, on his head or shoulders with Holy Scripture inside to represent the Law of God. From there you see a two-fold type of worship taking place. Many of the women fall to the ground as the Tabot passes by in reverence to the worship of God and the Ark it represents. Even though the feeling of holiness consumes the area as tens of thousands gather in white robes and prayer shawls, there is also a time of celebration with the chanting of prayers and the ever constant blast of "Lililililililili" and then an occasional "Sisisisi." As they march toward the makeshift tabernacle to place the Ark in the Holy of Holies for the evening of prayer and chanting, it brings you to an ancient time when the world was focused on God's presence on the Ark. A time when all people had a fear of the Lord, much unlike the times we live in today.

Secondly, I jump ahead to a time that is still to come, a time that Isaiah speaks about when the Ark of the Covenant will truly come out of its hiding place and the top of the Ark, the Mercy Seat, will be taken to the Holy of Holies in the third temple in Jerusalem to rest as the throne of Jesus.

"In that time shall the present be brought unto the LORD of hosts of a people scattered and peeled, and from a people terrible from their beginning hitherto; a nation meted out and trodden under foot, whose land the rivers have spoiled, to the place of the name of the LORD of hosts, the mount Zion." Isaiah 18:7

It will be a royal procession, much the same way that we see in Axum today during Timkat with the understanding that the ceremonial-adorned Levites will once again carry the Ark;

"From beyond the rivers of Ethiopia my suppliants, even the daughter of my dispersed, shall bring mine offering." Zephaniah 3:10

It will once again be a reverent celebration as the throne of God is brought before the King of Kings and the Lord of Lords.

Then I look at the type of worship that the people of Axum present during this ceremony. We see in 1 Thessalonians 5:17, *"Pray without ceasing,"* a charge given to all believers in worship to the one and only God, Jehovah. In Ethiopia I have never seen this statement presented with more truth than this. During the Timkat I witnessed men and women bow in prayer, many with their faces in the dirt, hours on end and even throughout the night never leaving that position on the ground. This made me think strongly about our worship and our prayer life. These people are sick, dying, and have very little in life but still thank the Lord in prayer for what

He has given them. We have far more than we deserve in life but yet our culture gives God just enough of our time to satisfy us. Our prayer revolves around our schedules instead of the other way around.

In all, it was like a symbolic belief but a real testimony of faith. These people were content with their lives. They seem to understand that their poverty and suffering are primarily because of the incredible task they have been given, to guard and protect the Ark of the Covenant. It makes you step back and ask yourself a question, why do these people claim to have the Ark, take so much time protecting something if they don't, and give all they have toward this Holy object? It really begins to make you wonder why monks spend their entire lives on a remote island with absolutely nothing there for them except to be the guardians of Holy ground that may have at one time been the location of a godly summit between the Father and the Son. Even though it all goes against the grain traditionally, I have to feel that their entire life's passion isn't spent for not. If we were given the task of protecting anything, I am confident that we would not have the passion or desire that these people have without time and a half pay, benefits, and extra breaks.

There is a real biblical feeling in this ancient land and one that cannot be dismissed. And then, when you consider the biblical references to what will one day take place there, it all begins to make a great deal of sense. God trusted these people to protect His secrets until a time when they would be needed again.

Confirming the Path

It's funny sometimes to see how God confirms your path. I strongly feel we lose so many blessings in our lives because we refuse to open our eyes and see the Lord work in us. I would like to share one quick example of how God continually confirms what He has laid before you. Months after returning from Ethiopia, the week had arrived for me to gather our team together and head down to the Bible publishing company to print the Ethiopian Amharic translations of the Bible for our return trip in January of the next year. Monday started off hectic, as always, with the basic chaos of getting the week ready for the numerous activities ahead. I had plopped into my seat in my office to finalize the covers for the Bibles from the photos that I had taken during our journey. A gentleman at the Bearing Precious Seed Bible Publishing Company had been working on the covers for a couple of weeks and was now waiting on my approval to finalize the project. They looked great and I sat in tears starring at the sample on my computer screen. To think that just months earlier I thought that I was heading out to search for the Ark of the Covenant, and now I'm writing a book and preparing to return to share the gospel with these wonderful people and to continue my research.

Tuesday was somewhat different. I had made my final calls to our team to remind them of Saturday's gathering to print and prepare the Bibles while carrying out a variety of other duties I needed to accomplish. Well, back in January I had put together a plan for a Bible study group that I also teach on Tuesday nights in our church's fellowship hall. Remember, this was planned in January and we are now talking the end of August. When I pulled up the study that I had prepared, which was the book of Acts that we had been on for several weeks with a couple of chapters that we were coming back to for further study, I realized that our final chapter was to go over, in more detail, Acts chapter eight with Phillip and his encounter with the Ethiopian Eunuch. My first thought was how fitting, we are going to print the Ethiopian Bibles on Saturday and today we just so happen to be concluding our study of Acts and returning to chapter eight, as I had preset the first week of the year. The study was amazing as I had a whole new perspective on why this man was in Jerusalem, and now in the desert, and puzzled by what he was attempting to read in the book of Isaiah. Bible study ended around eight o'clock this night, and Sherri and I went home to prepare ourselves for a good night's sleep. At ten I went up to our room and was about ready to shut off the lights, when my phone rang. On the other end of the line was my good friend Jonas, the Ethiopian who helped translate the Ge'ez Bible pages with Abba Yemanebirhan, in Cincinnati at the public library. He said, "Jim, do you remember I invite you to a celebration at our church when we meet?" I replied, "Of course, I remember." Actually, I had forgotten about it until he mentioned it and I just figured that it had already passed. He said, "I would like to ask you to come with your wife this Saturday to our church in Cincinnati in the morning." I was somewhat disappointed at first because I realized that we had to print the Bibles this Saturday morning. I regretfully told Jonas about our conflict and that we had to be there at nine in the morning, but he responded, again, with that assuring and bubbly Ethiopian response, "No problem!" He went on to say, "Our celebration and services begin at three thirty in the morning and you can come and be with us." I quickly said, "Three thirty in the morning, are you kidding me?" Jonas, thinking I was kidding said, "Yes, Jim and we will continue until after noon. You can come be with us?" I knew that I had been given the honor of being asked to attend this service and that it was the least I could do for what they both had done for me. I quickly said, "We will be there. And can I bring others with me?" He came back with what I expected, "Of course, you can. I see you Saturday morning!" He was excited and admittedly, I was too. It just hit me like a ton of bricks what was going on here. It was God's sign to me that He was confirming all that has and is about to happen regarding this journey.

Saturday arrived and the stars were bright in the early morning sky as we started out for our journey to the church in Cincinnati. We met up with Brad and Jenny Kiphart, a couple who felt the calling to be a part of our next expedition, and Jenny's sister Katy, as we loaded up and headed to the downtown services. I really had no idea where this place was, and I gave all my hope to the GPS to make sure we arrived in the correct location. As I thought, from my discussion at the library, this church was in the downtown area, but, much to my surprise it was in the rougher area of Over-the-Rhine, as it is known in Cincinnati. As we drove we came to a somewhat broken down location, but the GPS had us turn down a dark alley. I was beginning to wonder what I had gotten us into. But, somehow I knew that God was behind all of this, so there was nothing to worry about. We drove through the alley where there were three older buildings on the right and some homes that had seen better days on the left. I didn't see anything that reminded me of a church or anything that had anything moving around as if a church service was going on. So I drove our van out and went back around, turning back into the alley, for another pass. This time I rolled the windows down on both sides so I could listen. Then I heard it. Pounding slow rhythms on a large drum, or kebero, brought back the memories of the ceremony at Timkat in Axum as well as the bass pounding drum in the temple on Tana Kirkos Island. I pulled to the front of an old building with some type of covering over the door as we could barely see a speck of light shining through a tear in the covering. And then my hunch came true when I heard a loud call, "Lilililililililililili." This is the Ethiopian call in unison that we heard through the night, at a wedding ceremony near the Stelae Park, and at various other times while in Axum. I exclaimed, "We're in the right place."

Jesus: In Ethiopia

We parked in a small parking spot between two buildings and made our way in. Jonas just happened to be at the front door. He was thrilled we had made it. He had to step out for something so he made sure the others were aware we were there and they were instructed to help us in anything we needed. We removed our shoes before entering the sanctuary, and I noticed on the wall banners depicting Axum and Gondar. Most of the people, before entering the sanctuary would get down on their knees and bow low to kiss the floor, out of respect. The men and women were all wrapped in prayer shawls as the woman and children were on the right side of the isle and the men on the left. In the front was the Abba, along with the musicians on the drums and deacons or monks holding their prayer sticks and in unison, with a flick of the wrist, playing their sistrum, a silver or tin instrument likened to a hand-held combination of a small cymbal and castanet.

As we took our seats and watched, I couldn't help but wander back to Axum. My mind was flashing faces and places from this humble setting nearly eight thousand miles away. The Abba handed his prayer stick to another man as he took a break from the service. As he walked down the aisle, I made my way into the foyer of the church. He glanced once, and then took a double take when he saw my face. With a smile he gave a slight bow, as did I, and we exchanged the traditional Ethiopian handshake and went side to side toward each other in a show of respect. It was a comforting feeling, like I was at home and one with them.

We watched them worship hour by hour with chants of prayers and Scripture from their Bibles. Each word seemed to coordinate with the slow beat of the music. Occasionally others would make their way into the sanctuary, but at no time did anyone look at us any differently. Once in awhile, the same as in Axum, the drum beat would quicken and the Abba, along with the others of dignity in the front, would hit their prayer sticks on the floor while moving in closer to each other with one loud pounding of the kebero, and the prayer sticks hit in unison on the hard concrete floor covered in carpet. Just as the prayer sticks would hit the floor, the women would again give a loud burst of "Lililililililililili." It was amazing and was almost like a welcome home to the land of Ethiopia.

My watch ticked up to eight in the morning, and it was time for us to go. I had one of the men go to the front and tell Jonas to meet me in the foyer of the church. We shook hands and he leaned over to me and said, "Jim, remember, here you are always with friends." It brought tears to my eyes as I was reminded of the friendship and warmth of the people of Ethiopia. Upon returning from Ethiopia I had Larry Stinson, my jeweler friend from Tennessee, design some very unique pendants of the trumpets that we were honored to see from Moses in Axum's treasury. I reached into my pocket and grabbed a double silver trumpet pendant and gave it to Jonas as a gift of continued friendship. Tears began to show in his eyes as he nodded as if to say, "Thank you." We shook hands once more and as I turned for the door, the Abba was standing there. He reached for my hand and we both gave a slight bow in respect to each other. I told him I had to go and Jonas began to explain why we were unable to stay further into the service. It was an awesome experience, to relive the faithfulness of the Ethiopian people. But, for now it was time to go. We scurried to the van and loaded up, and I realized the importance of this relationship and how God directed us together.

We made it back to our church just in time to load everyone up and head to the Bible publishing building. Thirty people had gathered to help and the strong smell of incense was permeating from our clothes from the church service we had just attended as we entered into the van. Everyone was in a great mood and Sherri and I, the Kipharts and Katy were already on fire to get to work on the Bibles since our unforgettable early morning encounter. Before entering the publishing building, we all bowed our heads in prayer in the hopes that these Bibles will reach people all around the world. They say that the average Bible will be read by seven people. That figure gives me great hope to know that thousands will have their hands on these books and also that thousands more will be touched by what we are about to do.

The pages had been printed and the covers were ready to go. Our crew had to first assemble the Scripture in order and then put the appropriate covers over them. Since the pages were rough cuts, they had to be tapped down to get them as close to flush at the top as possible. With that complete we began to staple them together while others took the completed books by tens in an alternating manner. Then, five hundred or so at a time, we wheeled them out to others in our group, and they ran them through the cutter to make them into perfect Scriptures ready to box and take with us. I had pulled the first ones out of the stack to complete a particular project that Sherri had reminded me to do. Back in 2001, on a missions trip into Mexico, Dr. Verlis Collins, a missionary for this publishing company, was driving the bus that took our group into the country. As we drove for nearly eight hours through the Mexican desert, he shared his story about how people all over the world have their fingerprints on our lives. He said that someone led the preacher that shared the gospel of Christ with me to the Lord. So now, their fingerprints were on me. And that continues on and on. I had always remembered that from Mr. Collins. That always rang true for me anytime I had an opportunity to witness to anyone. I had brought a stamp pad with me and had all those involved stamp their fingerprints on the cover of this Bible. I later included John and Lisa Shoemaker's fingerprints on this cover as well. Although they were unable to be with us at the printing, they would play an important part getting them into Ethiopia on our next journey as will the Kipharts and the others on the team. I framed it and today that Bible hangs in my office as a reminder of our fingerprints on the world.

The Final Touch

I learned a lot while standing on the cliff and gazing into a vast desert alone with God. I'm a different person today because of what I have been shown, trusted with, and witnessed as God cleared a path. He presented me with an opportunity of a different sort in order to get my attention. He stopped me in my tracks and took away what pride I had to make me listen. He introduced me to a loving, humble, praying group of selfless people to show me how to trust only in Him as they do. Then when all was in order, He brought me to my knees, twice, to share a story about a meeting that took place long, long ago in a land far, far away from home. First, to call me with His trumpets to share a story with others to show how these trumpets will symbolize the calling out of His story to the world. Then, to my knees on a cliff side on an island shrouded in mystery and intrigue to give me the hidden journey of His Son and the meeting on an island flowing with righteousness. Even after I returned home, He continued to open doors and lay the crumbs along the path to finish His wish. I'm very different today than I was prior to this adventure. If anything else is true, I have learned to trust in God Almighty to direct my paths. If He is calling you to a place in your life, it's to your benefit to listen.

"*In all thy ways acknowledge him, and he shall direct thy paths.*" Proverbs 3:6

Jesus: In Ethiopia

I have learned through this adventure how true Solomon's wisdom really is. If we allow God to direct our paths then we really have so much more opened to us during our journeys through life. The words that my wife Sherri said to me just three days before we were invited to become part of this exploration have come true over and over again. As I thought in the beginning, I went on a journey to satisfy a desire believing I would become a self-made adventurer to gather clues for the lost Ark of the Covenant. I came away with a love and passion to help the people of Ethiopia and to share their need with others in a different sort of adventure. In the process of it all, God allowed me the privilege of uncovering one of His most hidden secrets and a revelation about His meeting with His only begotten Son on a remote island that was completely removed from the world.

This revelation of the mighty summit of God passing on to Jesus, His Son, the knowledge, power and grace to prepare Him for the greatest ministry in history may be one of the most revealing finds in recent years. For you and I this information puts an important piece in the puzzle to the missing years in the life of Jesus Christ and clarifies the one line of Scripture in Luke 2 to open our eyes to answer where and when Jesus received the wisdom and grace He carried with

Him. Although stepping back and looking at the whole picture now, I know that is only the beginning of our journeys through life. God allows us to travel through many valleys in our lives in order to prepare us for the great blessings that are still to come. In this adventure it is evident to me that we have the opportunity to make great friends through our continued travels to Ethiopia. The excitement builds each time we begin planning for the next journey to this remote land in Africa. Knowing that our fingerprints are on those Scriptures that people will be reading and consequently giving their lives to Christ, is reward enough for all the effort and preparation each time we journey there. It's also a thrill to know that God has so much more to reveal in this ancient land, and I can't help but believe that He will open more revelation to us with each and every journey there. But yet, in many more ways I still see the words that Sherri spoke becoming even more true in the time still ahead; *"Jim, I really feel that our biggest and best adventures are still to come."*

"When the trumpets sound…Arise and go!"

Reference Notes

1. The Holy Bible, King James Version
2. Robert Cornuke, *The Relic Quest* (Tyndale House Publishers, Inc.) 2005. (Various teachings of Robert Cornuke during our exploration can be found in this revealing book)
3. Flavius Josephus as translated by William Whiston, *The Complete Works: The Jewish War* (Kregel Publications) 1867, 1960
4. Answer In Genesis, *Uncovering The Real Nativity*, 2011
5. Paul Perry, *Jesus In Egypt* (Random House Publishing Group, A Ballatine Book) 2003.
6. Henry Sike, *First Gospel of the Infancy of Jesus Christ – First Translation*, 1697.
7. Jenny Jobbins, *Jesus in Egypt* (Al-Ahram Weekly On-line) 2001

8. Gabriella F. Scelta, *The Comparative Origin and Usage of the Ge'ez Writing System of Ethiopia*, 2008
9. Meera Lester, *Salome, Wife of Zebedee*, 2009
10. Lara Iskander and Jimmy Dunn, *An Overview of the Coptic Christians of Egypt* (touregypt.net) 2011
11. John Nash, *Archangel Uriel* (uriel.com) 2001
12. Mark Wingfield, *Journey to Bethlehem Must Have Been Grueling* (www.biblicalrecorder.org) 1997
13. Dinknesh Ethiopia Tour, *Gondar – Ethiopia History* (ethiopiatravel.com) 2011
14. Professor Sergew Habele Selassie, *The Establishment of the Ethiopian Church* (ethiopianorthodox.org)
15. Jerry W. Bird, *Barhir Dar, Lake Tana, Blue Nile Falls* (Africa-ata.org), 2010
16. Adamu Amare and Belaynesh Mikael, *The Role of the Ethiopian Orthodox Church in Literature and Art* (ethiopianorthodox.org)
17. Saint Takla Church, *The Christian Coptic Orthodox Church of Egypt* (Saint Takla Church Councils History, st-takla.org)
18. Dr. Zahi A. Hawass, *Holy Family in Egypt* (egypvoyager.com)

19. Ethiopiafamine.com, *Christianity in Ethiopia* (fhi.net, Food for the Hungry International)
20. *The Holy Family In Egypt* (holyfamilyegypt.com/map) 2007 Arab West Foundation
21. Wikipedia, *Kandake* (Wikipedia.org), 2011
22. Orthodoxwiki.org, 2011
23. Tibaldi, *Tribute to Orthodoxy: St. John the Theologian* (tributetoorthodoxy.com/theologian) 2009
24. Egyptian Holy Family Flight map, Chapter 5, www.touregypt.net
25. Ethiopia map, Chapter 6, www.mapsof.net

Jesus: In Ethiopia

Explorer: Bob Cornuke and Author Jim Rankin during the Jesus in Ethiopia exploration

Jim and Sherri Rankin at the Timkat ceremony in Axum

The Guardian of the Ark of the Covenant

Jesus: In Ethiopia

St. Mary of Zion Church in Axum, where they claim the Ark of the Covenant rests in today

Guide and Translator, Sesay, as he received Christ

Young Ethiopian couple at Stelae Park of the Kings

Timkat Ceremony as priests and monks carry a replica of the Ark through the streets of Axum

Jesus: In Ethiopia

Jonas and the Abba in Cincinnati, Ohio

The bathing pool of the Queen of Sheba at the Baptism of Christ ceremony during Timkat

Tadesse, a man learning to walk again because of the exploration team's passion for him

A boy and dinner for the family as our team purchased chickens for many in need

Jesus: In Ethiopia

A boy and his brother, they were never apart

Woman near death from tuberculosis and HIV

Marye, the Falashian Jewish woman on the mountain next to Gonder, Ethiopia

The view over ancient Axum

CPSIA information can be obtained at www.ICGtesting.com
Printed in the USA
BVOW081158190313

315837BV00001B/1/P